W9-ADO-479

HD
2,15
F63
994

FOCUSED QUALITY

Managing for Results

Harvey K. Brelin
Kimberly S. Davenport
Lyell P. Jennings
Paul F. Murphy

Foreword by
C. Jackson Grayson, Jr.

S^{t}L

St. Lucie Press

Copyright ©1994 by St. Lucie Press

All rights reserved. No part of this publication may be reproduced, stored in a retrieval system or transmitted in any form or by any means, electronic, mechanical, photocopying, recording or otherwise, without the prior written permission of the publisher.

Printed and bound in the U.S.A. Printed on acid-free paper.

Library of Congress Cataloging-in-Publication Data

Focused quality : managing for results / Harvey K. Brelin ... [et al.]
   : with a foreword by C. Jackson Grayson, Jr.
      p. cm.
   Includes bibliographical references and index.
   ISBN 1-884015-18-2 (alk. paper)
      1. Total quality management.   I. Brelin, Harvey K., 1942– .
HD62.15.F63   1994
658.5′62—dc20   94-17212
                                                              CIP

All rights reserved. Authorization to photocopy items for internal or personal use, or the personal or internal use of specific clients, is granted by St. Lucie Press, provided that $.50 per page photocopied is paid directly to Copyright Clearance Center, 27 Congress Street, Salem, MA 01970 USA. The fee code for users of the Transactional Reporting Service is ISBN 1-884015-18-2 8/94/$100/$.50. The fee is subject to change without notice. For organizations that have been granted a photocopy license by the CCC, a separate system of payment has been arranged.

The copyright owner's consent does not extend to copying for general distribution, for promotion, for creating new works, or for resale. Specific permission must be obtained from St. Lucie Press for such copying.

Direct all inquiries to St. Lucie Press, Inc., 100 E. Linton Blvd., Suite 403B, Delray Beach, Florida 33483.

Phone: (407) 274-9906
Fax: (407) 274-9927

Published by
St. Lucie Press
100 E. Linton Blvd., Suite 403B
Delray Beach, FL 33483

CONTENTS

APPENDICES

FOREWORD

This is definitely a book that you should read...that is if you want bottom-line results.

It was written because a growing—in fact, an alarming—number of organizations have gone down the path to total quality management, only to falter, fail, or worse, start drifting and blame TQM. Something is terribly wrong. Someone compared most TQM efforts to hunting ducks at midnight without a moon: there's a lot of squawking and shooting, but not many direct hits. That is why I encouraged the authors of this book to write it.

The American Productivity & Quality Center has been instrumental in generating a tidal wave of companies interested in improving quality. We were the first organization in the nation to begin serious work on what is now called the Malcolm Baldrige National Quality Award. We teach courses, hold conferences, and write articles on TQM. We consult on implementing TQM. In the process, we have found that some organizations have been very successful with TQM and benchmarking and have achieved bottom-line results. We've seen smiling CEOs and workers.

More often, however, we see what we call the "bewildered": those organizations that have trained everyone, but do not know how to use the training; those that have trained everyone in statistical process control, but do not know what to fix; and those that are making real progress in putting up inspirational posters, but do not know why they are not realizing much, if any, financial return on the dollars they have invested.

One thread runs through it all. Upper management either is or is not involved. When upper management does not drive the TQM effort, the results are disappointing at best.

Without giving the book away, I want to point out that the authors have experienced the same thing. In response, they have created a philosophy and a concrete methodology for producing results from TQM, which they call *focused quality management.*

It is critically important that you make TQM and benchmarking successful. It is important for your organization (whether you are in business, health, education, or government), for the people in your organization, and for everyone's standard of living. It may be a cliche to say that "We're in changing times, and we must meet global competition," but it's true. Those organizations that do not use customer-driven process improvement will not survive.

The bottom line is productivity. More productivity, in almost every case, leads to increased profits or improved effectiveness. The best way to get more productivity is through the use of quality concepts supplemented by benchmarking.

And the best way to get quality results is to read this book.

C. Jack Grayson
Chairman
American Productivity and Quality Center

PREFACE

Quality management can be either a winner or a waste. Winning quality efforts tend to follow the models of W. Edwards Deming and J.M. Juran, which pursue quality by improving the processes that result in products and services and, as a result, deliver concrete bottom-line impacts that convince employees, management, and shareholders that quality is the cornerstone of success.

Wasteful quality efforts, on the other hand, tend to tack on quality as an afterthought, relying on end-of-the-line inspections and nice-to-haves like quality circles, neither of which produces much quality overall. They also perpetuate the myth that high quality means high cost.

A method for focusing quality initiatives in order to achieve a winning edge is described throughout much of this book. Following the procedures described will help managers better satisfy customers, work more collaboratively with employees, and achieve high-quality results.

In Chapter 1, the importance of improving key business processes to achieve strategic business goals is highlighted. First, the vertical alignment of traditional organizations (the familiar top-to-bottom organization chart) is illustrated, in which little or no interaction across the structure is encouraged or, in worst cases, is even allowed.

Fundamental flaws are inherent in such a structure because most products and services require activity by many functional areas (sales, engineering, manufacturing, distribution, and the like) before they ever reach the customer. Thus, strict functional organizations simply do not work when products and services for customers are created in cross-functional processes.

To show how existing organizational structures evolved, Appendix

C traces the development of hierarchical, functionally driven organizations and explains why they are rapidly becoming obsolete as focused quality processes begin to dominate the workplace and marketplace.

Chapters 2 to 6 outline a process improvement approach that yields outstanding results as organizations put into place the elements of Focused Quality Management (FQM) to achieve strategic business objectives.

In Chapter 7, the critical role of the leaders of the quality initiative is described. Chapter 8 provides a review of key elements in the FQM process so that it can be determined whether or not an organization is prepared to take the quality journey.

The final chapter lays out the most common causes for quality initiative failures and offers helpful hints to solve the problems.

ACKNOWLEDGMENTS

The authors have many people to thank for their help in creating the concepts and techniques discussed in this book. First and foremost we wish to thank our clients for giving us the opportunity to learn and grow with them. We also owe a debt to our colleagues Jerry A. Fuller, Ronald R. Malone, Susan T. Siferd, and Brian J. Andrews for helping to shape our *prepare–plan–deploy–transition* approach, which is the heart of Focused Quality Management. The quality graphics job done by Fred Bobovnyk of the American Productivity & Quality Center speaks for itself.

We owe a special thanks to Dr. Carla O'Dell for her contributions on benchmarking. Lastly, we owe Dr. C. Jackson Grayson an immeasurable tribute for his vision, his innovativeness, and his guidance and support, without which we would lack the knowledge and the drive needed to complete this effort. To all who have helped us, we say thank you.

AUTHORS

Harvey K. Brelin is President, American Productivity & Quality Center. Mr. Brelin has extensive experience in manufacturing and service industries managing improvement efforts and providing training and motivational presentations.

Prior to joining the American Productivity & Quality Center (APQC), he was Vice President, Quality, at First Bank in Minneapolis and held quality-related positions with Alcoa of Australia, Clark Equipment Company, and Ford Motor Company.

Mr. Brelin holds B.S. and M.S. degrees in Mechanical Engineering from Wayne State University. He is an ASQC-certified quality engineer and a certified reliability engineer.

Kimberly S. Davenport is a Consultant, APQC-Consulting Group. Ms. Davenport has extensive experience working with AT&T in Washington, D.C. and in support of the federal government. One of her major initiatives at AT&T was to use total quality management principles to reduce the development and delivery time of product and services to federal customers.

Ms. Davenport holds a Bachelor's degree in Journalism and a Master's in Organizational Development. She is currently pursuing a Ph.D. in Human and Organizational Development at the Fielding Institute.

Lyell P. Jennings is President and CEO, APQC-Consulting Group. Mr. Jennings has 30 years of senior corporate management and consulting experience. He has focused on consulting for process improvements through participatory management and process improvement teams.

His activities have been global in nature, with international consulting experience in both Europe and Latin America.

He previously served as regional partner in a Big Six consulting firm, providing training and implementation services in the areas of operations improvement, total quality management, and information systems.

Mr. Jennings has a Bachelor of Science in Economics from the University of Illinois and a Master of Business Administration from the Rochester Institute of Technology.

Paul F. Murphy is Executive Vice President, APQC-Consulting Group. Dr. Murphy has extensive experience implementing quality improvement in the Department of Defense, industry, and government and has consulted and instructed in total quality management with a variety of clients, large and small, U.S. and international.

He previously served in the White House as Special Assistant to the Director, Office of Management and Budget (OMB) and Military Assistant to the Secretary of Defense and later served as Chairman of the Department of Command and Management at the Air University, where he received the Air Force's highest award for contributions to Professional Military Education.

Dr. Murphy received his Bachelor of Science degree from the U.S. Naval Academy and has an M.B.A. and a Ph.D. in Business Administration.

Part I

MANAGING
THE VITAL FEW

In Chapter 1, the foundation is laid for the basic premise upon which this book rests. The reason why many total quality management efforts have failed to achieve the results management desired is outlined. Simply stated, many failed efforts set out with bright hopes for serene sailing to the promised land of higher quality, lower cost, and increased customer satisfaction without a map, a compass, or a rudder. As a result, the high hopes of the ships of quality foundered on the reefs of limited resources, management inattention, and uninspired leadership. The results were predictable. At best, the voyage found islands of excellence in a sea of opportunities. At worst, the crew abandoned ship, saying, "I knew it wouldn't work here."

The approach recommended in Chapter 1 calls for a clear set of sailing guides. It addresses the need for focused quality—that is, quality improvement in those key processes that have an impact on the vital few things: the Critical Success Factors (CSFs) that **must** be done to achieve the strategic objectives of the organization. The idea is simple, yet its application can be difficult. To be successful in their efforts, managers must take the helm and steer the quality effort. As the chapter points out, managers who do this with a good compass—**a clear vision**—will have smooth sailing in their quality voyage.

1

WHY FOCUSED QUALITY MANAGEMENT?

Quality isn't something you lay on...like tinsel on a Christmas tree. Real quality must be the source...the cone from which the tree must start.

Zen and the Art of Motorcycle Maintenance

WHERE TQM FALLS SHORT

At a conference, 150 business executives were asked, "How many of you have been involved with some kind of quality effort?" All present raised their hands.

Then they were asked, "How many of you are satisfied with the results?" Absolutely no one responded.

Although this is a small statistical sample, it nonetheless sums up what many people currently think about Total Quality Management (TQM): It does not work. Yet a whole industry exists to provide quality training, quality consulting, process improvement, process redesign, and process reengineering. The White House talks about rein-

> "Failure to improve process performance results in failure to improve organization performance. Failure to effectively manage processes is failure to effectively manage the business."
>
> Geary Rummler and Alan Brache[1]
> *Improving Performance:*
> *How to Manage the White*
> *Space on the Organization Chart*

venting government. The Commerce Department administers a National Quality Award. The president has a quality award for government agencies. States and companies are creating their own quality awards. A subindustry is aggressively marketing a plethora of prizes, pens, and pins stamped with a quality logo. Is all this activity a waste?

From the volume of articles, bulletins, and complaints, it is obvious that too many organizations are spending too much money and too much effort, all in the name of quality, and getting little in return. TQM, as a management discipline, is taking the blame for the failure to get results.

The authors' ten years of experience with over one hundred organizations, providing training and consulting support to their quality implementation, benchmarking, and reengineering efforts, has revealed some impressive successes. However, some organizations fall far short of their improvement goals.

Most of these failures occurred because the organizations did not integrate their TQM efforts with their management practices to achieve strategic **business objectives.** As a result, TQM became the responsibility of the human resources department or some other staff function and was disconnected from the operational management mainstream.

Focused Quality Management (FQM) is an approach intended to prevent such results. The basic premise of FQM is that quality initiatives must be targeted at improving those processes that have the greatest impact on what must occur if the organization is to achieve its objectives. These **critical success factors,** or CSFs, are the essentials, and the **key processes** that impact them should be the focus of management attention; all else is secondary. The key processes typically are cross-functional in scope, involving a number of departments or divisions in the delivery of products and services to the customers of the organization. Successful quality initiatives recognize this fact and structure their improvement processes and teams accordingly.

Other factors also impact the success or failure of the quality initiative. The manner in which the organization communicates, trains, rewards, and measures are systemic means that should enable it to consistently carry out the key processes. The organization and its employees should never lose sight of the fact that they are the means, not the ends. The key processes are, and must remain, the focus of the improvement initiative.

Focused process improvement is what it takes for quality to have positive results. What this means is that work gets done (or does not get done) and organizations succeed (or fail) based on what occurs within specific key business processes. By linking strategic goals, quality efforts, the business processes that most affect customers, and critical success factors, FQM does work, and organizations get results.

As mentioned earlier, many organizations find themselves cut, bruised, and on the edge of abandoning TQM. Fundamentally, their quality efforts have failed because they were not focused on improving a strategic business process. Instead, quality teams were left to drift and ended up working on such trivial matters as what color to paint the cafeteria.

Not unusual is the CEO who asked for help after investing thousands of hours training all the company's workers and not getting any results. When asked what the people are doing, he replied, "We have 450 quality teams!" "Yes, but what are they doing?" he was asked again. His answer said it all: "I don't know."

When the workers were asked what they were supposed to be doing, their response was, "We don't know." They said they met once a week for two hours and talked about how important quality was. They got a free baseball hat for showing up.

Unfortunately, this is not a unique story. Many organizations share the same kind of aimless, "feel good," "have a hat" unfocused quality programs. Managers are becoming as frustrated as the workers are with TQM initiatives that promise much but yield little.

There are three fundamental reasons why TQM efforts miss the target and do not accomplish important business goals. First, they are not focused, meaning that TQM is being implemented sporadically, with little regard for what is important.

Second, senior executives are not intimately involved in managing quality. This might be called the "bubble-up theory." Executives send managers and employees to training, create a lot of teams, and wait for a host of ideas and improvements to bubble up. The executives think that quality, customer satisfaction, and cycle time will automatically improve and that they will realize lower costs and higher profits. They are like parents who want their kids to have a good education, but do not take the time to read with them at home. It does not work that way. Bubble up is all fizz and no focus.

The third reason is that some organizations have vision, but lack the discipline needed to make it come true. Their TQM efforts flounder because they do not:

- Focus on the processes important to strategic goals
- Include senior executives in the selection process to identify specific improvement projects
- Provide for ongoing executive review to ensure that implementation produces both short- and long-term results

Evidence clearly shows that a disciplined, management-led, focused approach is essential if quality efforts are to succeed. The approach must be targeted to improve those vital few **cross-functional business processes** that result in a product or service for customer use.

Focus is what distinguishes championship athletes, dominant sports teams, and world-class organizations. In a business, a manager creates that focus.

Of course, the elimination of all vertical structures is unrealistic at best and an invitation to anarchy at worst. Instead, products and services are delivered to customers through a series of interrelated organizational processes, whether it be in business, education, or healthcare (see Figure 1.1). No single manager or group of operating personnel has sole ownership of the process—except for the CEO, but they all share the responsibility for its success or failure.

Functional Organization

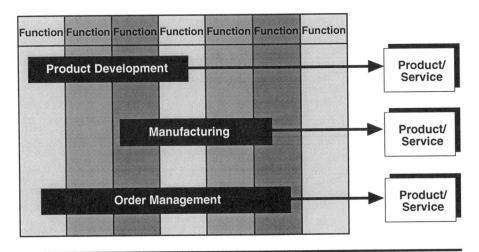

FIGURE 1.1

Thus, the key business processes of an organization must be managed across all the functional areas that play roles in delivering products or services. What are key business processes? The answer to that question is largely determined by the manager and the type of business.

The important thing to realize is that key business processes are the heart of the business. Identifying and improving them will in turn realize dramatic improvements in the quality of products, services, and the **bottom line.**

THE WAY IT IS

In a perfect world, all workers could envision the business process of which they are a part. They would create their work, coordinate activities, manage interfaces, be very proactive, quickly deliver products or services, and produce high-quality work.

The world, however, is not perfect. Almost all organizations are structured and managed through functional departments (for example, sales, marketing, administration, engineering, manufacturing/operations, distribution, customer service).

Many business organizations resemble the hierarchy shown in Figure 1.2. Within each functional department are well-educated, well-trained, and highly motivated managers and workers—good people who know the business. Such a functional structure effectively reduces a complex environment to manageable units. However, functional structures ignore the fact that products and services are delivered to customers through horizontal processes, that is, key business processes that cross functional department lines and require cooperation from many departments.

In every organization, even if one department has primary responsibility for production, others play roles, from marketing to warehousing. That is the real world. Unfortunately, most organizations and most quality efforts pretend that the world looks like the organization chart. The results are predictable—not good, but very predictable.

A MAZE OF FRUSTRATION

To illustrate this concept, imagine a patient going into the hospital for minor surgery. During his stay, many different hospital departments

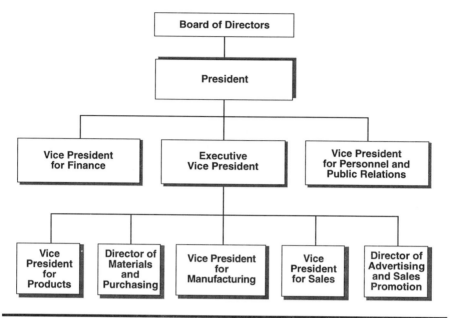

FIGURE 1.2

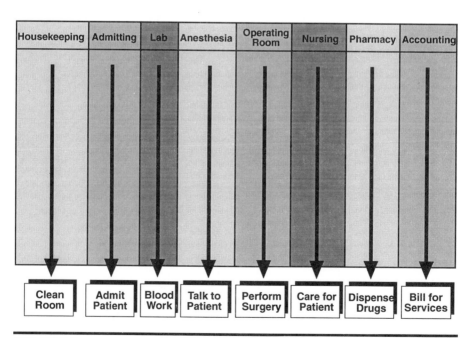

FIGURE 1.3

are involved in his care. Each delivers a product or service, as the chart in Figure 1.3 illustrates.

It appears as if each functional department were producing a separate product or service, and they may well be. However, those are simply functional products or services that are parts of one end product: the patient's recovery.

The problem with most functionally focused TQM efforts is that they try to improve functional products and services without fully addressing how and how much each contributes to the ultimate product. In this example, hospital admissions may do its job well and produce a superb admissions product. The nursing staff is top-notch. The operating room's surgeons are the best to be had anywhere.

Nonetheless, in functionally focused organizations things can fall through the cracks. Admitting processed the paperwork efficiently, but forgot to tell housekeeping that the patient needed a clean room and a freshly made bed. The result was that the patient had to wait.

The lab technician expertly and painlessly drew the patient's blood, but did not know that his next stop should have been anesthesiology for a pre-op interview with the anesthetist. The result was that the patient went back to admitting, several floors away, only to be sent back to anesthesiology, next door to the lab.

What this hospital has are some functionally superb people who are unaware of the total process. No one is responsible for the crossover points between functions. Each department sees its own product, but no one has the big picture.

Naturally, functional structures do serve their purpose. There is a real need to do admissions correctly, surgery correctly, and the like. Moreover, functional structures tend to simplify staffing decisions, allowing senior managers to effectively assign personnel and streamline operations.

Although it would seem that the benefits of such measures would enhance an organization's ability to serve its customers, in reality they do not. What happens in the real world looks more like a maze, which is what happened to the patient's hospital stay, as illustrated in Figure 1.4.

After making his way through the maze, the patient was frustrated and unhappy, yet each function may have believed that they treated him well. In the case of a hospital stay, patients may have little choice and suffer in silence. However, in many cases (including healthcare) consumers are showing an increased willingness to challenge such

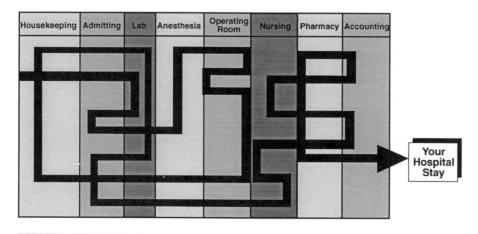

FIGURE 1.4

inefficiency and poor service. They are also rebelling against the costs of this approach.

Work thrown over the fence at the transition points not only annoys and inconveniences customers, but can lead to a lot of rework. In fact, rework has been shown to eat up 30 to 50% of the total cost of doing business. Just as important is the impact on current and potential customers when products or services are late, wrong, or too expensive.

The key point to remember is that customers care about and react to the end product that is the sum of all the functional products. For example, customers visiting Disney World care little about what goes on behind the scenes to create the magic. A hotel guest should not be concerned with the process that gets the right room clean and ready for a stay. A frequent flyer remembers on-time arrivals, not the fuel load or pilot schedules. Consumers instinctively focus on the ultimate outcome they desire.

Yet over and over again in the workplace that focus is lost. Sales worries about sales, engineering worries about engineering, purchasing guards its territory. Manufacturing merrily makes things. Customer service fixes problems, and executives execute plans or people, depending on how it goes. Once in a while everything comes together. All too often, however, everything falls apart, and customers lose. When customers lose, the business loses them. That is the inevitable result of what can only be termed serendipity management.

THE WAY IT SHOULD BE

The chart in Figure 1.5 shows how it should be: key business processes crossing smoothly through functional departments toward common goals. The way to achieve these results is through Focused Quality Management.

Focusing quality efforts puts the emphasis on managing key business processes across the organization. Focused quality manages cross-functional processes rather than taking a functional approach that matches the organization chart.

In other words, FQM reflects the way that products and services are delivered to meet customer needs. By focusing on cross-functional processes, rather than functional outcomes, the probability of meeting strategic business goals and satisfying customers with higher quality at lower costs is increased.

Further, FQM takes responsibility for process improvement out of human resources or organizational development and puts it in management, where managers direct the outcome. FQM requires that executives and senior managers get involved and stay involved—that is, if they want results.

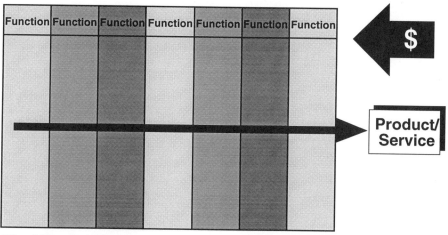

Company/Organization

The Way it "Should Be"

FIGURE 1.5

FQM is a viable management discipline that should, can, and does work when properly implemented. "Properly implemented" means that **managers** must lead the effort.

FQM puts the manager where he or she belongs: in the strategic leadership role. The manager defines the organization's strategic goals. The manager identifies the key business processes. The manager selects the process improvement strategies and projects and monitors progress to make sure the key business processes are structured and managed so as to achieve the strategic goals.

> "**W**e didn't put Xerox through a major restructuring for the fun of it. We did it because we had to. It was change or die."
>
> David Kearns
> Former Chairman and CEO
> Xerox Corporation

Managers do not sit around wondering what went wrong with TQM or why it works in Japan and not in the United States. In a wide variety of organizations, executives and senior managers have demonstrated that FQM can turn a struggling organization into a winner and lead a successful organization to market dominance. The rest of this book explains how they are doing it.

BIBLIOGRAPHY

1. Rummler, Geary A. and Alan P. Brache, *Improving Performance: How to Manage the White Space on the Organization Chart,* San Francisco: Jossey–Bass, 1990.

Part II

FOCUSED QUALITY MANAGEMENT: A FRAMEWORK FOR SUCCESS

In Part II, the foundation that must be laid before a manager can improve key business processes and make quality an all-pervasive part of the organization is described. Reading it will help to identify critical success factors and assess key business processes. A number of analytical tools—from process flow charts to process capability analysis—are also presented and explained. These tools can aid in better understanding an organization and how to improve it.

In particular, a proven four-step approach for implementing Focused Quality Management is described in detail. The steps in this approach are called *prepare, plan, deploy,* and *transition.* The words are not as important as what they represent: necessary steps along the way to improving key business processes.

Each organization is unique and must take advantage of the good things that it has already accomplished. Therefore, this book does not offer a cookbook methodology. Instead, the approach is easily tailored to make it extremely useful in helping to ensure that

■ Process improvement is connected to strategic business goals

■ The focus is on improving business processes instead of pursuing non-integrated projects

■ Upper and middle managers are aware, committed, and managing the improvement process

■ The effort builds on existing achievements

■ The entire organization assumes ownership of the improvement process

WHY THE EMPHASIS ON FOCUS?

Tightly focusing quality initiatives can and will produce results. Focus is sharpening thinking to a fine point consistent with organizational objectives. First, the vision is very broad. Strategic goals are tighter, but still broad. Identifying critical success factors and the key business processes that affect them begins to get more specific. That focusing and refinement is the key to success.

For example, perhaps "effectively communicating with customers" has been identified as a critical success factor. In a highly competitive, price-sensitive marketplace (say, telephone long-distance service), communicating with customers promptly and effectively about billing questions or notifying them when there is a dramatic difference in their calling patterns (perhaps a calling card has been stolen) are but a few ways that communicating with customers can make a difference. Thus, in this instance, the focus narrows in on the customer service process and looking for ways to improve how inquiries are handled and how employees proactively communicate with customers.

In the end, the focus should be on the things that are most important to the business, but that are not being done well. The focus should *not* be on the things that waste time and effort, because they are of no real value to the ultimate strategic goals of the organization (see Figure II.1).

HOW DO YOU GET FOCUSED?

The first step is *preparation*. This is where the manager and the leadership team get organized and develop statements of vision, mission, and values. Key processes are identified and an assessment is

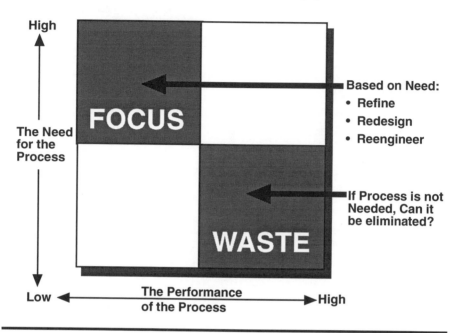

FIGURE II.1

conducted to determine both where the company is and where it wants to go. The next step involves *planning*. This is where tangible plans are developed for improving the key business processes and the organization as a whole. This leads to the all-important step of *deploying* process improvement teams to carry out the improvement process. The final step is the *transition* to instituting Focused Quality Management as the way business is done every day. This approach to implementation is depicted in Figure II.2.

It is not simple, but it will be successful as long as managers commit to investing time and effort. Not everyone who has realized a successful quality improvement effort has used this approach. However, every organization that has used this approach has achieved success in improving the quality of its business processes, service to its customers, and bottom-line results.

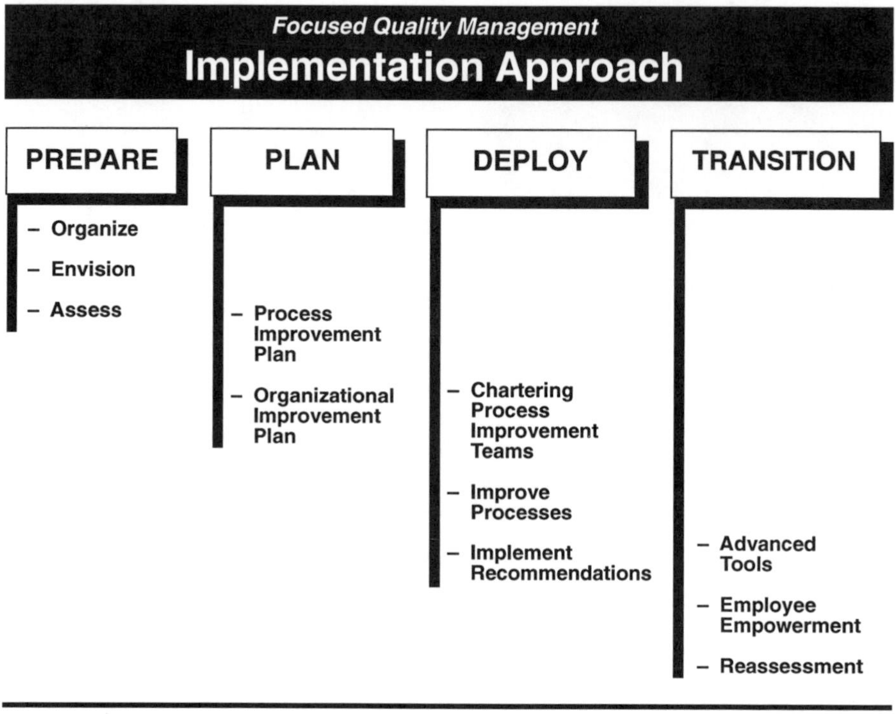

Focused Quality Management
Implementation Approach

PREPARE	PLAN	DEPLOY	TRANSITION

- Organize
- Envision
- Assess

    - Process Improvement Plan

    - Organizational Improvement Plan

        - Chartering Process Improvement Teams

        - Improve Processes

        - Implement Recommendations

            - Advanced Tools

            - Employee Empowerment

            - Reassessment

FIGURE II.2

2

PREPARE:
LAYING THE
FOUNDATION

Although customer satisfaction may seem an obvious goal, it is not usually achieved. Why? Is it intentional? Probably not. People usually come to work wanting to do the right thing. They also want to do things right the first time, and they continually strive to improve so they can satisfy customers.

Many organizations conspicuously post quality goals such as "Do the right things," "Do things right the first time," "Strive for continuous improvement," and "Satisfy customers." If no one comes to work planning to make mistakes and make customers angry, why does it happen so often? The answer may be that often executives have not successfully focused the organization.

Most employees think they are doing what management wants them to do. It would seem, then, that management has a problem

This Conversation Will Never Happen

Question: "What do you plan to do today?"

Answer: "Well, I hoped to do the wrong things."

Question: "All day?"

Answer: "Nope, only until 10:30. Then I will correct my errors until lunch."

Question: "What about after lunch?"

Answer: "Then I'll try to make things worse around here."

Question: "All afternoon?"

Answer: "No, only until 3 p.m. Then I'll call some customers and annoy them."

effectively communicating what they want employees to do. Is this lack of communication intentional? Again, probably not. Ineffective communication most likely occurs because executives lack a clear vision for the organization. Therefore, they cannot clearly articulate what they expect from their employees. As a result, the organization lacks direction. As Dorothy was told in the land of Oz: "If you don't know where you're going, any road will do."

Only top management can, and should, set the course. That is why management must get involved early on in the quality process, beginning with crafting the mission, vision, and values statements of the organization.

Prepare is the first phase in the focusing process. It requires a real commitment from management because, more often than not, the organization must change its thinking and put customer satisfaction at the top of its list of priorities. Everything else is secondary. As Tom Peters puts it in his presentations, "If you satisfy customers, you'll make a lot of money."[2]

If quality is put first, the entire organizational culture will reflect that thinking. In order for that change to occur, however, management must believe in and effectively communicate the focus on customer satisfaction throughout the organization.

> "Indeed, some Japanese firms, and the quality experts whose precepts they follow, have emphasized process over result. To the quality faithful, if the process is managed, the result takes care of itself. This philosophy seems to have worked well in Japan, and we are reluctant to raise the specter of national character differences. Yet, in the American business culture—noted for emphasis on large, exciting change and visible results—a process orientation that shuns results thinking may meet substantial resistance."
>
> Thomas H. Davenport[1]
> *Process Innovation*

The *prepare* phase of the focusing process is comprised of three main steps (see Figure 2.1):

- Organizing
- Envisioning
- Assessing

ORGANIZE A QUALITY LEADERSHIP TEAM

In order for quality to become part of an organization's culture, the team that leads everything else important to the organization must be

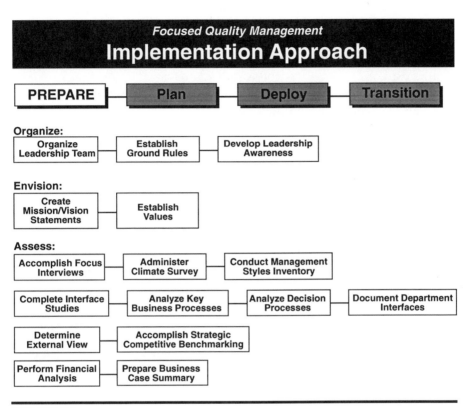

Focused Quality Management
Implementation Approach

| PREPARE | Plan | Deploy | Transition |

Organize:

| Organize Leadership Team | Establish Ground Rules | Develop Leadership Awareness |

Envision:

| Create Mission/Vision Statements | Establish Values |

Assess:

| Accomplish Focus Interviews | Administer Climate Survey | Conduct Management Styles Inventory |

| Complete Interface Studies | Analyze Key Business Processes | Analyze Decision Processes | Document Department Interfaces |

| Determine External View | Accomplish Strategic Competitive Benchmarking |

| Perform Financial Analysis | Prepare Business Case Summary |

FIGURE 2.1

committed to quality. To begin with, the executive staff must organize into a quality leadership team. This means the CEO and his or her direct reports—*the people who run the business.* They control budgets, allocate resources, and make day-to-day decisions about the enterprise.

This team must make quality the organization's way of life by integrating it into every decision process. Cost and schedule remain important; however, when quality comes first, it is amazing how much gets done right, on schedule, under budget, and meets customer expectations.

Just as responsibility for the bottom line cannot be delegated, neither can responsibility for quality be delegated. Instead, managers must take personal responsi-

"The new impetus for quality will be limited only by the pace at which our CEO's accept responsibility for their roles, which cannot be delegated."

J.M. Juran[3]

bility for managing quality. Quality must be examined in a broader sense and its strategic implications recognized.

A number of firms have found leadership awareness seminars to be useful, especially two-day, intensive, hands-on training sessions. For example, during the seminar the executives get to perform as a team in a business environment. Having been given what seems to be a simple task, the first attempt is always an eye opener for the executives. In the hundreds of times this exercise has been attempted, every team loses money and generates only about 40% acceptable quality. Once they see the dismal results and realize that they have failed using their traditional management approach, the executives are willing and anxious to learn. They now know why the team needs to understand *focused quality management* and the value of quality concepts.

An important characteristic of the seminars is that they are conducted by experienced consultants who are able to make the connection from the academic theory to the real world.

During awareness seminars, consultants teach the benefits of cross-functional process improvement, quality and reengineering principles, benchmarking, organizational structure, roles and resource requirements, and quality tools. For example, they discuss the seven simple process analysis tools of quality improvement: histograms, Pareto diagrams, run charts, scatter diagrams, control charts, flow charts, and cause-and-effect diagrams. This allows the quality leadership team to start to use a process improvement focus, rather than the problem-solving focus that is so often a part of corporate culture.

> "**A**s the quality crisis deepened, U.S. companies launched initiatives to improve their quality. Most failed because of ignorance…For decades, these managers had delegated quality to a quality department, never acquiring the training and experience needed to set appropriate quality goals and develop a plan of action for achieving them."
>
> J.M. Juran[3]

The effect of this approach becomes clear when the executives try the same task again after learning the process improvement concepts and tools. The results are always impressive. Every team makes money and as a rule produces 100% quality. The results convince the skeptics, and the resultant buy-in is a powerful force for change.

After training, the Quality Leadership Team (QLT) usually sets an appropriate roll-out strategy for the quality initiative and establishes ground rules for the QLT's activities. Those rules must clearly tell everyone watching that nothing is more important to the QLT than quality.

For example, the QLT must demonstrate that every member believes that the only way to better serve customers is to improve business processes. To do so, the team must work collaboratively, serving as an organizational role model for the cross-functional teamwork required for process improvement. For people reared in a functional organization, this is not an easy task. Yet this ability to "walk the talk" is a key determinant of an organization's potential for success. It is also closely watched by everyone in the organization, since they are hoping to pick up clues as to the behavior expected of them.

ENVISION THE ORGANIZATION

Once the QLT is established, the next step is to clearly define the mission, vision, and values of the organization.

The **mission statement** (see Figure 2.2) describes the organization's work clearly, concisely, and concretely. It says what the company does. The mission statement should be relevant and easily understood by everyone, from the CEO to the person in the mailroom. It should be customer-oriented, focusing on the products and services provided, *not* the things done to get them. The mission statement lays the foundation

---

Mission:
The organization's reason for existing

☑ Describes the organization's work clearly, concisely, and concretely

☑ Is relevant and clearly understood by everyone

☑ Should be customer-oriented, focusing on products and services

☑ Lays the foundation for other design and direction components

☑ Provides direction and a sense of purpose for all

FIGURE 2.2

for everything that follows. If done well, it provides direction and a sense of purpose for all. A great deal of thought should be invested in the mission statement, because it outlines what is being done and sets the stage for where the organization is going.

Domino's Pizza had a well-done mission statement. Simple but clear, it stated that Domino's aims to "deliver a hot, quality pizza in less than 30 minutes at a fair price and a reasonable profit."

Mission statements—if done well—sound so simple, but how difficult are they to create? Kevin Campbell, president of ServComp (a privately held company specializing in developing, installing, and maintaining mission-critical, enterprise-wide information systems for mid-sized companies), said that the company's mission statement seems rather simple:

> ServComp specializes in developing and installing and maintaining mission-critical, enterprise-wide information systems.[4]

However, as Campbell was quick to point out:

> This...current mission statement...has gone through a number of iterations. The first time we went through a quality envisioning exercise, we developed a mission statement that was similar; but as we've grown and as we've evolved, so has our mission statement.

Sometimes, the issue is how to develop a mission statement for a large, diverse organization. Ned Sickle, vice president of Marriott Corporation, said:

> Creating a mission statement is a real challenge for a large company like Marriott because we are in lots of different businesses. We are in the hotel business, as many people understand, with our Fairfield Residence, Courtyard, and full-service properties. We also have a large services business: Marriott Management Services, with food services and facilities; Host, which is airport food service for airlines and airport cafeterias; and senior living services. So for us, the challenge is to have mission, vision, and values statements that really are business specific.[5]

The **vision statement** (see Figure 2.3) depicts the kind of organization that a company wants to become or how the organization is to be seen or thought of by customers, employees, and others. It sets the direction that everyone works toward, especially when procedures and

---

Vision:
The organization's future, ideal state

☑ **Depicts the kind of organization that you want to become or how you want to be seen or remembered**

☑ **Sets the direction that everyone works toward**

☑ **Empowers people and creates enthusiasm by highlighting the organization's distinctive contributions**

☑ **Affords a basis for identifying gaps between the current state and the future state**

FIGURE 2.3

rules do not apply and when difficult decisions must be made. It empowers people and creates enthusiasm by highlighting the organization's distinctive contributions that employees can feel pride in making. The vision statement also affords a basis for identifying gaps between the current state and the envisioned future state.

At a Chamber of Commerce Quality Conference, Dr. Paul Murphy of the American Productivity and Quality Center spoke about one organization's vision:

> Possibly the best example that I've seen of vision was when my son was in the Disney summer program. That's a program where college students work at Disney Land or Disney World during the summer for three hours' college credit.
>
> I went down to see him just about a week after he started, and I couldn't believe it. He was saying things like "Excuse me." "How can I help you?" "Have a great day." Well, I had him 21 years and tried to teach him that. Disney had him one week and here he was practicing such courtesies. Even though he was not on duty, he was constantly looking around to see if everything was okay. I was impressed and I told him so.

He told me: "Dad, this is supposed to be the happiest place on earth, and that's why all of us who are cast members here at Disney have to make sure that our guests are indeed happy, because if you have one sad face walking around this place, it only takes a few minutes before other people are looking sad, and by the end of an hour a lot of people in the park could be walking around frowning. So we have to make sure that doesn't happen, and if we see it happening, we have to fix it."

I said, "How did you learn that?"

He said, "Oh, they give us a lot of training. We had a class Monday."

I said, "Well, how do you know exactly what to do?"

He said, "Well, you just follow Walt's vision to make this the happiest place on earth, because that's the way Walt wanted it to be."

I told him, "Robert, I don't know how to break this to you—Walt's dead."

He said, "Not here."

He was right. It was obvious everywhere I looked. At Disney, Walt's vision is alive, and all the employees understand it in their heads and in their hearts. That's what makes Disney special. That's why we pay so much to be part of the magic.

With the recent emphasis on quality, it is all the rage for executives to go away for a weekend retreat and come back with a vision statement. But how do you take that vision statement and make it meaningful for all employees? How do you make the vision statement change the way they behave?

According to Ned Sickle of Marriott:

That's one of the real challenges, because oftentimes the senior executives will come away from a weekend retreat with mission, vision, and values statements; and although they provide an overall framework, they don't really identify specifically what behaviors are expected from an employee. What we've done at Marriott Corporation is to take the vision and values a few steps further in terms of more detail.

For example, *integrity*, *balance*, and *customer-driven* are all elements of the vision and values for Marriott Management Services. I know, as an employee, that when we talk about *balance* that means, for example, that I can expect to take vacations that I have already scheduled. That's a very specific behavior. Another example is senior executives are evaluated by their peers and subordinates on how well they live up to our vision and values. Those are just a few examples of how we take mission, vision, and values statements and make them tangible in terms of expected behaviors.[5]

Think of the best in any field. What probably comes to mind is a man or woman who was able to see a future, ideal state and get others to share that vision. The results can be impressive. In medicine, in business, and in government, vision and the ability to let others see it is what sets the real leaders above the rest.

What also sets them apart is the understanding that the vision must be meaningful and motivational to those it is supposed to guide and motivate. That understanding leads them to question the clarity of the vision and be willing to test it. For example, a CEO and the members of his staff took their draft vision statement into the halls of their building and asked people at random to tell them what it meant. The results were enlightening. The CEO summed it up with, "Well, gang, I guess it's back to the drawing board." In the end they crafted a much better vision statement that everyone could understand.

A key step in the envisioning process is developing the **values statement** (see Figure 2.4). The values statement is the beacon by which everyone should navigate. It establishes the organization's

---

Values:
The organization's code of ethics

☑ Establishes the organization's basic beliefs, the principles that guide the corporate culture

☑ Centers everyone on a shared behavior model

☑ Emphasizes customer satisfaction, quality management, continuous improvement, employee empowerment, and cooperative relationships

FIGURE 2.4

beliefs, which are the principles that guide the organization in its day-to-day life. The values statement centers on those things that, when truly believed, focus everyone on a shared behavior model. Therefore, the values statement emphasizes customer satisfaction, quality, continuous improvement, employee empowerment, and cooperative relationships.

The values statement, when done well, is a document that inspires people. It makes them proud to be part of the organization. It serves as a personal benchmark, guiding employees and shaping behavior. For example, at A.C. Nielsen company's Nielsen Marketing Research division, the values statement crafted four decades ago by the founder (Figure 2.5) still is looked to as the standard against which today's activities are measured. The values that shaped the company's past are used today to mold its future. The fact that Nielsen people still keep this statement on their office walls makes it a timeless example of effective communication.

Often, an organization's advertising reveals its values. For example:

Value	Advertising
Innovation	*Just slightly ahead of our time* (Panasonic)
Customer service	*We get paid for results* (Cigna)
High performance	*The ultimate driving machine* (BMW)
Customer relationships	*Your business partner* (IBM)
Reliability	*It keeps on going, and going, and going... the Energizer* (Energizer Batteries)
Quality	*Quality is job 1* (Ford)

Sometimes, an organization's values may seem to conflict. Then, it becomes necessary to clearly establish the priorities and the overarching values. For example, valuing both innovation and quality might create a conflict when the company tries to be first with industry breakthroughs. Although the product may be the latest and greatest high-tech widget, if it does not consistently perform, customers soon perceive the company as one that does not produce high-quality, reliable products. Thus, the value of being first to the marketplace might need to take a back seat to achieving and keeping a reputation as a producer of reliable, high-quality products.

It is easier for employees and managers to establish priorities when

The Nielsen Code

IMPARTIALITY
Be influenced by nothing but your client's interest.
Tell him the truth.

THOROUGHNESS
Accept business only at a price permitting thoroughness.
Then do a thorough job, regardless of cost to us.

ACCURACY
Watch every detail that affects the accuracy of your work.

INTEGRITY
Keep the problems of clients and prospects confidential.
Divulge information only with their consent.

ECONOMY
Employ every economy consistent with thoroughness,
accuracy and reliability.

PRICE
Quote prices that will yield a fair profit.

DELIVERY
Give your client the earliest delivery consistent with quality—
whatever the inconvenience to us.

SERVICE
Leave no stone unturned to help your client realize
maximum profits from his investment.

A.C. Nielsen

Chairman, A.C. Nielsen Company

Nielsen

FIGURE 2.5

they understand the overarching values. They are forced to choose often, and an explicit statement of values can help them choose well.

The vision, mission, and values then are the guiding light that illuminates the path to the future. Done well, envisioning lets us see where we are going—**the vision**. It also clearly establishes our course—**the mission**—and the way in which the organization will function—**the values**. Taken together, these things keep us focused on our goal, our course, and our actions as we strive to improve. They help us keep sight of and integrate our long-range, short-range, and day-to-day activities.

ASSESS THE ORGANIZATION

The most important information that the organization receives during the *prepare* step is revealed when the following are assessed:

- ▪ Customer satisfaction
- ▪ Process performance
- ▪ Employee attitudes

The quality assessment is designed to show clearly (through the use of data) what is working in the organization and what is not. Using that information, additional data can be gathered to find out which business processes most influence customer satisfaction. That knowledge, in turn, helps focus the improvement effort on those few vital factors that most impact strategic goals. Assessing the current state also provides a baseline for measuring progress, as the company strives to fix those vital few core processes that need improvement.

Further, an organizational quality assessment helps to identify exactly what is taking place within each area and function as work is done. Because the assessment itself is process-focused, it also makes people more aware of and focused on processes.

Finally, the assessment includes a great deal of emphasis on measuring how well customers are served, because everything should connect back to satisfying their needs.

Before beginning the assessment, the QLT reviews the mission and vision in order to identify specific objectives and critical success factors essential to achieving the vision and strategic business objectives. They then identify the processes that have the greatest impact and target the bulk of their assessment activities on those, thus using a disciplined approach to zero in on the key processes.

To begin, existing organizational objectives are reviewed for clarity and consistency with the vision and mission of the organization. The mission is what the company does, the vision is where it is going, and the objectives are the things that must be done to get there. Objectives sharply focus the mission and vision statements by providing tangible measures to let the company know how it is progressing on its

"The very act of planning requires differentiating between the important and the unimportant and establishing a hierarchy of objectives."

Charles Ferguson and
Roger Dickinson[6]

charted course. For example, if the vision is industry leadership, the objectives may be to increase sales revenues and volume by 20% in 1994, or if the vision is innovative responsiveness, the objective might be to reduce cycle time on product development by 30% by year end. In essence, objectives are those things that can be done and measured so that the company knows how it is progressing toward its goal.

Next, critical success factors (CSFs) are identified for the organization. CSFs are those few things that are absolutely essential to success in achieving objectives.

By way of background, the term *critical success factors* for many years was used in a corporate planning context to mean the most important subgoals of a business, business unit, or project. The definition is modified here to mean those things an organization must do to meet objectives and achieve its vision. Identifying CSFs is absolutely essential because it asks: "What actions and results are imperative to achieve our mission?" and "How are we going to do it?"

Senior managers are often surprised at what relatively mundane and unglamorous things turn out to be critical to success. Identifying CSFs drives home the concept that excellent organizations achieve excellence by doing little things excellently.

> "The thing about these critical success factors is that they are incredibly common things. What we determined is that if we could do these common things uncommonly well, we could be very successful."
>
> Kevin Campbell[4]
> Founder and President, ServComp
> Chamber of Commerce
> Quality Conference

For example, one of the critical success factors at Marriott is accurate client billing. Marriott Management Services has about 3000 different food and facilities accounts spread out across the country in hospitals, colleges and universities, and schools, as well as corporate accounts that submit accounting information to its processing center in Buffalo. The Buffalo service center processes that information and then produces client statements. Ned Sickle, vice president at Marriott Corporation, explains:

> Those statements have to be accurate, and we recognize the importance of that accuracy. What we found was that a significant proportion of our client statements weren't as precise as we wanted them to be. To address this problem we used one of the quality tools called Pareto which is a graphical tool of ranking

causes of problems from most significant to least significant. In our case we formed a small cross-functional process improvement team, and found that about 10% of our accounts were involved in 90% of the billing complaints. So we were able to focus in on those accounts, provide some training to those unit managers, and as a result we were able to improve the accuracy of our client statements.[5]

Lest you get the impression that the Marriott experience is unique, consider the words of ServComp founder and president Kevin Campbell:

When we first went through the quality program, we sought input from our customers and identified five critical success factors. Those five critical success factors were: number one, we needed to educate our customers about how we deliver our service; secondly, we needed to respond quickly and effectively when they call for service; thirdly, we needed to arrive on site according to our contractual commitments; fourthly, we needed to promptly and accurately fix problems; and what absolutely blew me away was the fifth critical success factor, which was billing them accurately. This was a very important issue. The thing about these critical success factors for us is that they are incredibly common things. What we determined is that if we could do these common things uncommonly well, we could be very successful. As a result of focusing on these, we grew, over a five year period, about 771 percent and we made the *Inc.* 500 list in 1991.[4]

Here are two companies in different industries, one a corporate giant, the other a small enterprise, one managed by thousands, the other by fewer than a dozen. Both found a relatively simple and unnoticed process to be a key to success. Both found that out by asking their customers!

Doing common things uncommonly well is an important point when identifying what an organization's CSFs should be. Just as important, or even more so, is to use the customer input gathered during the assessment to learn what specific factors are most important to customers. That list should be the

"The critical success factor concept is very simple: For every organization, there are between three and eight things that are critical."

John F. Rockart[7]

nucleus of the CSFs and should drive the development of the final list. For example, if customers indicate that responsiveness to problems is important, then there should probably be a CSF focusing on reduced cycle time in identifying and resolving service problems.

CSFs also need to be flexible and evolutionary. As time goes by, certain CSFs may move up and down the priority list due to economic or external factors. For example, in a tight economy, something not so important in the past, such as payback, may suddenly escalate to the top of the list, because being able to survive in the short run is the only way in which the vision could ever be achieved over the long run. However, CSFs themselves do not change significantly, and usually come back to the same things in terms of customer satisfaction, responsiveness, accuracy, and similar concepts.

CSFs need to be based on consensus among the members of the QLT. Coming to an agreement on a list of absolute essentials is crucial in gaining a sincere commitment and willingness on the part of the management team to do whatever it takes to achieve an organization's mission and vision. When that has been accomplished, the results speak for themselves.

A final point about CSFs is the need to select only those few and truly critical goals that will make the difference to success. Things such as an ultra-modern information system may be nice, but if it is purchased and the company still fails, was it really critical? Remember, the goal is *focused,* not *fragmented,* quality management.

A powerful technique for ensuring that the CSFs include only those few and vital things is an acid test of sorts called *necessary and sufficient.* Once a final list of CSFs is generated, the QLT revisits the mission and vision statements and asks: "Does this list of CSFs cover only those things *necessary*

"In one study, the top 10 managers in 125 European companies were asked individually to identify their companies' 5 most critical objectives. The minimum number from each company would be 5; the maximum, 50. Managers of the most profitable companies agreed on 6 to 12 objectives. For the 40 worst companies, the range was 26 to 43. In other words, the top executives of the poor performers had no shared vision of what they were trying to do, while just the opposite was true of the successful companies' leaders. Significantly, a few years after the managers of one worst category company had agreed on its critical objectives, the company moved into the most profitable category."

Maurice Hardaker and Bryan K. Ward[8]
"Getting things done"
Harvard Business Review

(absolute musts) *and sufficient* (if it happens, the objectives are met) for the attainment of the mission and vision?" If not, the discussion and refinement of the CSFs should be repeated until there is clear agreement that the list covers all and only those areas *necessary and sufficient* to achieve company goals.

However, just trying to implement quality by managing CSFs is not enough because CSFs are outcomes. They are not the "how to" of an organization; therefore, they are not directly manageable.

What is necessary is to focus in on the key business processes that are going to make the CSFs come true. To do this, the QLT must identify which processes have the greatest impact on the CSFs. They also must make judgments regarding the current performance of the process. The goal is to identify those processes which, when improved, will have the highest potential payoff. A handy tool to assist in this analysis is the **process prioritization matrix** (Figure 2.6).

Most organizations do not have the resources to simultaneously fix each and every process. They therefore must decide which processes should be looked at now and which processes can be addressed in the future. The QLT must strike a balance when allocating resources necessary for simply running the business and those used to improve it. A process prioritization matrix can help. To illustrate how this matrix technique can be used to help prioritize processes for improvement, let's take a look at how it was used by an investment company (see Figure 2.7).

First, the investment company listed its key processes and CSFs as shown. The impact each process had on each of the CSFs was rated in the matrix as 1, 2, or 3. A score of 1 means the process had little or no impact on the CSF. A score of 3 means the process had an essential impact on achieving that particular CSF. The total impact of each process on the critical factors was added and placed in the "total impact" column. The process performance score from 1 to 10 for each key process, as determined earlier during the assessment, was placed in the next column to the right. The next column shows the process performance gap, which is the process performance score subtracted from 10. The "weighted gap" column is the product of the process performance gap score multiplied by the total impact. The last column simply shows the ranking of the weighted gaps of the processes from highest to lowest.

The QLT determined that they would focus their assessment on the MIS, distribution, and record-keeping processes This does not mean

Process Prioritization Matrix

	Critical Success Factors						Total Impact	Process Performance	Process Performance Gap	Weighted Gap	Priority
Key Processes											
1											
2											
3											
4											
5											
6											
7											
8											
9											
10											

Rating Key:

Process Impact on CSF:	Process Performance:
1 = Low	1 = Inadequate
2 = Medium	5 = OK
3 = High	9 = Very Well

FIGURE 2.6

Process Prioritization Matrix

Rating Key:

Process Impact on CSF:	Process Performance:
1 = Low	1 = Inadequate
2 = Medium	5 = OK
3 = High	9 = Very Well

	Critical Success Factors										
Key Processes	PROFIT ON INVESTMENTS	RESPONSIVE TO CUSTOMERS	INTERNATIONAL MARKET PRESENCE	ASSET SECURITY	COST EFFECTIVE OPERATIONS	SKILLED, MOTIVATED WORK FORCE	Total Impact	Process Performance	Process Performance Gap #	Weighted Gap	Priority
1 MARKETING	1	2	3	1	1	1	9	7	10-7 = 3	27	
2 SALES	1	2	3	1	3	1	11	8	10-8 = 2	22	
3 INVESTMENT ANALYSIS	3	1	2	3	3	1	13	8	10-8 = 2	26	
4 RECORD KEEPING	1	3	1	2	2	1	10	6	10-6 = 4	40	3
5 CUSTOMER SERVICE	1	3	2	1	1	1	9	7	10-7 = 3	27	
6 PERSONNEL SELECTION	2	2	1	2	2	3	12	9	10-9 = 1	12	
7 DISTRIBUTION & MAILING	1	3	2	3	2	1	12	5	10-5 = 5	60	2
8 MANAGEMENT INFO SYSTEM	3	3	1	2	2	1	12	4	10-4 = 6	72	1
9											
10											

\# 10 = PERFECT PROCESS

FIGURE 2.7

that they ignored the other processes; the processes still must be managed to the best of the firm's ability. It is simply a matter of attempting to focus on what is important and in need of improvement.

THE PROCESS ASSESSMENT

The assessment is a rigorous investigation of key processes to determine what is being done, where it is done, and how it is done. The value added and cost of each activity is measured, time lines are established, and step, subprocess, and process cycle times are recorded and analyzed to determine waiting time versus working time.

Rework and scrap are identified, measured, and costed. The financial impact of the process is measured, and a detailed comprehensive business case is built to show the potential return on investment of changes to the process, as well as the costs of making the changes. In the case shown in Figure 2.8, process improvements were identified which would yield savings of $1.8 million while improving accuracy and customer satisfaction. The assessment report is a product of:

■ Interviews with managers to identify the shortcomings of the key processes

■ Focus group interviews with process operators

■ Customer surveys and/or analysis of existing customer data

■ Internal surveys

■ Macro-level process analysis to identify flow, cycle time, cost, bottlenecks, suppliers, and customers

As part of the assessment, the team prepares analyses of cost, inventory, cash flow, and waste. The final output is a comprehensive list of opportunities and their costs and benefits (Figure 2.9).

Identification of process improvement opportunities and benefits is only part of the assessment process. The assessment team also looks at the organization's communication, training, reward and recognition, and measurement systems. Their purpose is to identify barriers to improvement, as well as those things the organization does well or can do to assist in making Focused Quality Management (FQM) part of the organization's culture. Combined with the process improvements identified during the process assessments, these organizational improvements make the quality initiative a reality by putting in place the

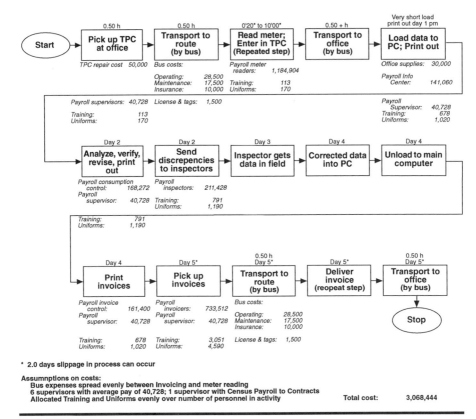

FIGURE 2.8

processes and day-to day procedures that make quality a way of doing business.

Upon completion of this part of the assessment, the company will have a much better focus and understanding of what needs to be done during the next step in the FQM process.

Because people are the key to success, the assessment also looks at attitudes and perceptions.

Part of an assessment can be a quality systems survey. In the survey, employee perceptions of the organization's quality practices are determined. The structure for gathering, analyzing, and reporting employee input is often a set of widely known criteria, such as that used for the Malcolm Baldrige National Quality Award. Those criteria have evolved

Recommended Action

Plans are underway to restructure routes and improve the workload balance for meter readers and invoice drivers. Replacement machines for the TPC's are also being evaluated. However, this process has so many flaws that a Re-engineering Team should be assigned to completely redesign it.

Benefits

Re-eengineering this process would:

☐ Reduce costs by approximately $1.462 million annually;

☐ Reduce invoicing cycle time and accelerate cash flow with a present value of approximately $400,000 annually;

☐ Improve accuracy and thus reduce problems in downstream processes such as Complaints; and

☐ Improve customer satisfaction

Assumptions underlying cost saving estimates are:

☐ *Meter readers:*	read meters quarterly and invoice average consumptionin other months; reduce head count by 50%	
	Payroll, Uniform and Training Savings	$598,936
☐ *Inspectors:*	reduced reading frequency reduces discrepencies; reduce head count by 67%	
	Payroll, Uniform and Training Savings	$142,984
☐ *Consumption Control:*	reduce head count from 6 to 1 through automation	
	Payroll, Uniform and Training Savings	$141,060
☐ *Invoice Deliverers:*	rebalance workloads and assign 2 routes each; reduce head count by 50%	
	Payroll, Uniform and Training Savings	$340,576
☐ *Information Center:*	work simplification; reduce head count by 20%	
	Payroll, Uniform and Training Savings	$28,552
☐ *Invoice Control:*	work simplification; reduce head count by 20%	
	Payroll, Uniform and Training Savings	$32,560
☐ *Supervisors:*	reduce proportionate to other payroll deuctions (50%)	$122,184
☐ *TPC's:*	replace; reduce maintenance cost by 50%	$25,000
	Total Annual Savings	$1,461,512

If average consumption values are used for invoicing two months out of every three, then two-thirds of all invoiced amounts could be delivered on day 1 of the invoicing cycle rather than day 5. The effect on cash flow would be:

☐ Annual amount invoiced	$176,660,000
☐ Proportion affected by accelerated invoicing	2/3
☐ Amount accelerated	117,773,000
☐ Daily time value of money *	.00085
☐ Present value of cash flow per day	100,017
☐ Days cycle time reduction	4
Annual Benefit	$400,428

* Based on 22% annual interest and 9% annual inflation

A cost profile for the meter reading and invoicing work groups is shown on the next page.

FIGURE 2.9

over time and now represent a practical and competitively focused approach to evaluating an organization.

Tom Sanders and his management team at Shell Oil Company's Lubricants Division used the Baldrige criteria to evaluate their quality performance. The assessment identified a need for greater customer focus and stronger internal capabilities. Major information system needs were identified, and customers were brought in to assist with the design of the new system. Today, Shell Lubricants is implementing a state-of-the-art system that provides customers with 24-hour on-request access to technical and commercial information, as well as automated order fulfillment. Additionally, sales force automation and internal management reporting will improve overall operating efficiency and effectiveness.

> "**Q**uality for quality's sake was a fad. Quality to make a difference is the quality wave of the future."
>
> Tom Sanders[9]

Recent changes to the Baldrige criteria reflect the focus. The award is being recast as a qualifying award (a company that meets the standard receives the award), rather than as a competitive one (e.g., "You're good, but companies A and B were better. Sorry.")

It is also filling a gap by allowing nonprofit healthcare and educational institutions to apply for the award. These organizations will compete in their respective categories by 1995 at the earliest. Next, government agencies will be allowed to participate.

Since 1988 the award's focus has steadily shifted from form to substance. Results became increasingly significant and activities in pursuit of the results less important in the grading process. This reflected a shift in the attitudes of business organizations. After a brief period of measuring quality effectiveness by the magnitude of the difference in activities, the focus shifted to the current approach of measuring the difference in results.

The emphasis on results has done a lot to focus the quality movement and in fact is likely to increase its long-term viability. Experience shows that businesses will use and support what works for them. As long as business and management believe that FQM and process improvement make a difference, they will support the initiatives in pursuit of strategic business objectives.

The recent changes in the Baldrige criteria make them a more useful, though still limited, standard against which an organization's performance can be compared to that of organizations with similar processes.

After the quality systems survey is complete, the organization has a much better idea of how well it stacked up against the criteria, the industry, and the world. Why is that information important? It provides a benchmark for comparing the organization's performance with others, especially with quality award winners—in other words, proven achievers of the excellence for which the company is striving. It shows where the company is keeping pace and where it is falling short. It often identifies immediate opportunities.

For example, as mentioned earlier, the assessment includes a quality climate survey and focus group interviews with employees to ascertain their perceptions regarding their processes. It also includes studies and analyses of process and departmental interfaces (the "white space" on the organization chart), in-depth analyses of the steps performed in the key processes, a review of decision points, and processes employed to make the decisions.[11]

Armed with this data, an assessment team can also conduct strategic and competitive benchmarking to determine how much improvement is needed and/or possible. As mentioned earlier, the improvement opportunities are then subjected to rigorous financial analyses to determine the cost and benefits so that a results-focused business case summary can be prepared. That summary is a fundamental input to the decisions made in the *planning* steps that

> "**L**ike many proud U.S. companies facing hard times, Cummins Engine Company was an eager convert to the gospel of quality. In 1983, as the king of truck engines reeled from recession and foreign competition, its executives made a pilgrimage to Japan. Back home, they installed a formal quality process. It didn't work. Problems soared on a new engine, warranty costs doubled from 1987 to 1989, and customers deserted. So in 1991, the 72-year-old Columbus (Ind.) company turned a harsh spotlight on itself. It began to judge its performance using the criteria of the Malcolm Baldrige National Quality Award. It even applied for the award, just to get feedback from Baldrige examiners.
>
> "It didn't come close to winning. But the feedback prompted it to pinpoint the source of trucker complaints, train workers better, and seek help from quality leaders like Xerox Corp. Cummins now finds production defects in fewer than 1 percent of its engines versus 10 percent before. Its warranty costs are down more than 20 percent since 1989. And in 1992, it expects to emerge from three years of red ink."
>
> John Carey, Robert Neff, and Lois Therrien[10]
> "The prize and the passion"
> *Newsweek,* 1991

follow. The business case explains the likely benefits, as well as costs of the improvement, while integrating this perspective into the overall strategic plan. As a result, a company may decide to improve a process based on strategic issues that would not warrant improvement when reviewed solely on a short-term financial basis.

In summary, the major part of the assessment is focused on helping to identify how CSFs are impacted by major processes and to measure how well they are performing. To do that, it is important to interview the people who are truly the experts—the employees who actually do the work. If there are holes in the process, these are the people who will know where they are.

Asking the right questions will elicit their insights into the good and bad things about the major processes and just how effective and efficient those processes are. Combine that with careful analysis and a clear picture of where things stand should emerge. With that information, it should be possible to target improvements to the key business processes, which will lead to achieving strategic business goals.

The assessment of key processes provides the knowledge required to determine if the existing processes need to be refined, redesigned, or fundamentally reengineered—the "3 R's" of process improvement. It helps the quality team to understand what change is needed, how much is required, and where on the process improvement continuum (Figure 2.10) they should focus their efforts.

Last, but not least, it depicts how well customers are being satisfied and to what extent they are satisfied with the specifics of the company's performance. To do this, a company must evaluate a variety of historical and current sources of information it may already have about customer satisfaction. If necessary, the company may also interview customers to make sure their requirements are completely understood.

At this point in the process, the company has identified:

- What it does (the mission)
- Where it is going (the vision)
- How it will behave (the values)
- Where it is (the assessment)
- Who is going to lead the way (the QLT)

The QLT has a clear idea of what can and must be done. As shown in Figure 2.11, the result of the assessment effort is the ability to draw

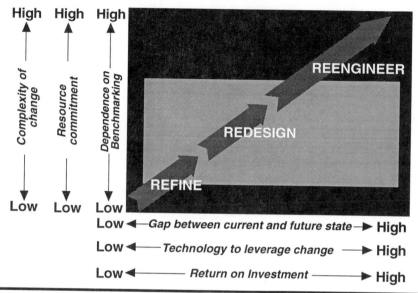

FIGURE 2.10

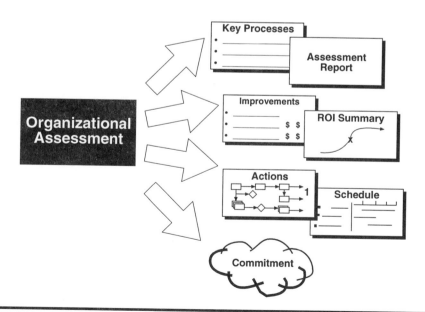

FIGURE 2.11

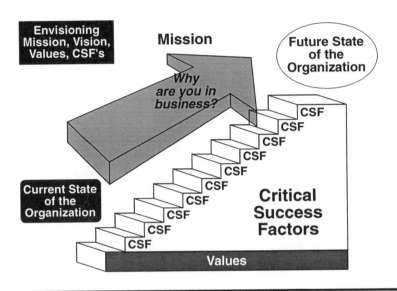

FIGURE 2.12

accurate conclusions, identify strategic opportunities, and create an improvement approach based on return on investment that assures buy-in from everyone involved.

In the *prepare* phase, the foundation has been laid for the rest of the quality process (see Figure 2.12). With a firm foundation in place, it is time to build a FQM plan based on what has been learned. How to create a successful FQM plan is outlined in Chapter 3.

BIBLIOGRAPHY

1. Davenport, Thomas H., *Process Innovation: Reengineering Work through Information Technology,* Boston: Harvard Business School Press, 1993.
2. Peters, Tom, "Passion for Customers," Video Publishing House, Inc, 1987.
3. Juran, J.M., "What Japan Taught Us about Quality," *The Washington Post,* pp. H1 and H6, August 15, 1993.
4. Campbell, Kevin, "Getting Started in Quality Management," Chamber of Commerce Seminar, Spring 1993.
5. Sickle, Ned, "Getting Started in Quality Management," Chamber of Commerce Seminar, Spring 1993.
6. Ferguson, Charles R. and Roger Dickinson, "Critical Success Factors for Directors in the Eighties," *Business Horizons,* p. 16, May–June 1982.

7. Rockart, John F., "Critical Success Factors: A Method to Help Managers with Information Planning," *The Consultant,* p. 1, Nov.–Dec. 1984.

8. Hardaker, Maurice and Bryan K. Ward, "Getting Things Done," *Harvard Business Review,* p. 114, Nov.–Dec. 1987.

9. Sanders, Tom, personal communications, Spring 1994.

10. Carey, John, Robert Neff, and Lois Therrien, "The Prize and the Passion," *Newsweek,* p. 59, Winter 1991

11. Rummler, Geary A. and Alan P. Brache, *Improving Performance: How to Manage the White Space on the Organization Chart,* San Francisco: Jossey-Bass, 1990.

3

PLAN:
FOCUSING PROCESS
IMPROVEMENTS

The second step in Focused Quality Management (FQM) (see Figure 3.1) is the creation of a strategic quality plan based on the results of the organizational assessment previously described. The strategic quality plan has two components: a process improvement plan, so those key

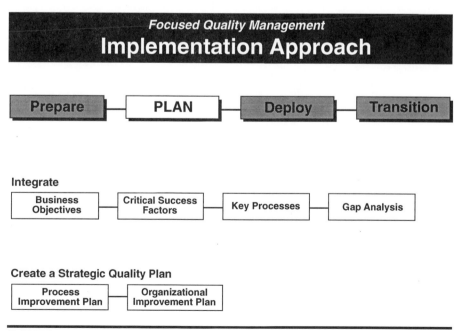

FIGURE 3.1

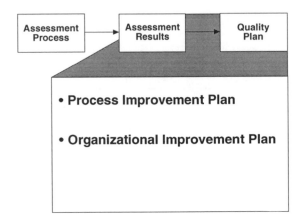

FIGURE 3.2

processes that are important to company strategy and the customers can be improved, and an organizational improvement plan, so the future focused quality organization can take shape (Figure 3.2).

The strategic quality plan formalizes actions and priorities to achieve the mission, vision, and business objectives; communicates priorities and provides direction to all employees; and provides an organizational road map for FQM.

In this chapter, the overall concept of the first part of the *plan* step, the process improvement plan, is described. A methodology that has been used very effectively by a number of industry and government organizations to accomplish their planning is then discussed in greater detail.

The process improvement plan is based on the assessment and is designed to close the gap between current and future states. To create a process improvement plan, a consensus must be reached on: (1) **business objectives** (what the organization wants to achieve), (2) **critical success factors** (what must happen if the organization is to achieve its objectives, and (3) **key business processes** (what most influences the ability of the organization to achieve its goals while effectively satisfying customers.

The results of the key process assessments are used to develop improvement strategies. These strategies are used to select process improvement opportunities, against which teams are chartered to re-fine, redesign, or reengineer projects in order to improve them by

Strategic Quality/Business Planning

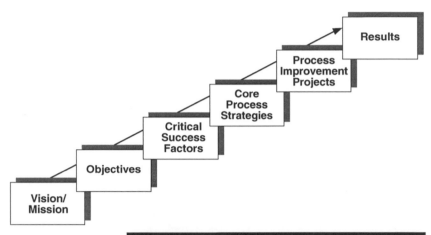

To move from current state to future state requires a significant change in the way business is conducted. Business is conducted through processes.

FIGURE 3.3

eliminating non-value-added steps and closing gaps in performance. The process improvement implementation plan reflects all the decisions that have been agreed upon by the Quality Leadership Team (QLT) during the *prepare* and *plan* phases.

Figure 3.3 depicts how the very broad and, by necessity, abstract organizational vision and mission created earlier were specifically focused and made concrete by integrating them with the objectives, critical success factors (CSFs), and key processes that most impact the achievement of the objectives. During this planning phase, the culmination of this focusing effort is the selection of process improvement projects that directly support the organization's strategic goals and objectives. This ensures that the projects are targeted at meaningful results on the one hand and, because of this, secure the necessary continuing management support on the other.

This completes the overview of the *plan* phase. Let's now take a more in-depth look at the specific steps used to create a process improvement plan.

Typically, during this process, the QLT participates in a planning session to select process improvement projects that are consistent with the vision, mission, goals, and objectives of the organization. Its goal is to develop a time-phased plan to deploy process improvement initiatives based on the results of the organizational assessment. The vision and mission statements, along with the process assessments and customer satisfaction data, and knowledge of the CSFs are used as a compass for selecting improvement projects and targeting specific short-, mid-, and long-term opportunities for improvement. In essence, the QLT designs its entire quality implementation program so that it links the overall vision, mission, and objectives of the organization with the specific day-to-day activities being carried out.

In order to do this, the QLT reviews the results of the assessment to see if they confirm their earlier judgments regarding the importance and performance of the key processes. In most cases, the assessment confirms their appraisal and the QLT uses a very simple but powerful technique called **gap analysis** to select the processes that will be the focus of their improvement efforts.

In performing the gap analysis, the QLT references the process data that were collected during the assessment. This information is evaluated to determine the need for the process in terms of its relative importance in meeting the organization's goals and how well it is being performed.

The QLT scores each process in terms of performance and importance using a scale of 1 to 10 (10 is high and 1 is low). For example, a member of a QLT at a large hospital has just assessed its surgery process and found that surgery is canceled or rescheduled 12% of the time because the lab results are not available in the patient's medical records. After discussion among the QLT members, she agrees to rate the process as a 3 in terms of performance.

Next, the key processes are ranked against each other to determine their relative importance to the organization's overall success. For example, the team member gives the surgery process a high score of 10 because it is perceived by the QLT as the most important. The scores for both the performance and importance of the key processes can then be plotted on a scatter diagram (Figure 3.4).

Highly important processes that are performed well fall in the upper right quadrant (e.g., the happy face quadrant). Since it makes little sense to spend a lot of time or money fixing something that is already done well, the effort should focus on the upper left quadrant, which contains processes that are highly important but not done very well. Unless the

Capability Gap Analysis

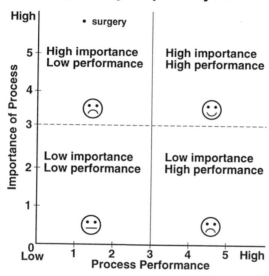

FIGURE 3.4

process can be improved, the goals of the company will not be achieved. The lower right quadrant also shows an unhappy face, because these processes are not being done very well, but they are not very important. This means that limited resources, which could be used to help improve the processes that are in the high-importance, low-performance quadrant, are being wasted.

Using a capability gap analysis helps the organization to focus on the most important processes. If improved, they will have the greatest influence in helping the organization succeed. The assessment data are used to help identify specific projects within the processes, so that teams can be chartered to carry out the improvement process.

As a client astutely observed, "Wasting resources—being better than required at unimportant things, while important things lack adequate resources—makes me feel dumber than dirt!"

Another tool many organizations find extremely useful for evaluating key processes is **benchmarking.** In benchmarking, process results are compared to the results attained by other organizations that perform the same process. Benchmarking can increase the QLT's awareness of what a really good process can achieve and usually results in a lower but more valid assessment of their own process.

These results tell the QLT how big the gap is between where they are and where others are. Thus, using benchmarking for gap identification can help executives gain perspective on the performance strengths and weaknesses of their strategic processes and how their results compare to those of other organizations.

At this point in the FQM process, the QLT has achieved clarity and agreement on the mission and future vision of the organization. The team also knows, based on its organizational assessment, how well things are getting done, along with what is important to customers. With all this information, it is possible to determine which processes are high impact and need improvement. A lot has been accomplished.

Armed with this information, including a clear understanding of the potential return on investment, the company can rank its process improvement initiatives in order of priority. The assessment results have guided efforts throughout this phase, and the plan can now be put on paper. The process improvement plan acts as a road map for the initiatives. Like all maps, it shows the way to a desired destination. The QLT, however, must continue to manage progress once the path to take has been chosen.

After this analysis, the next question is how the organization will make the necessary improvements. Here, the QLT selects process improvement projects and prepares charters to guide the teams as they work on them. Some "rule-of-thumb" criteria for selecting process improvement projects include:

- **Cost:** Can it be done given the resources available?
- **Return on investment:** Will the benefits be worth the costs?
- **Visibility:** Will it demonstrate tangible results?
- **Time frame:** Can it be accomplished in a reasonable time?
- **Long vs. short term:** Is there a mix of quick and longer term projects?
- **Difficulty:** Can it be done?
- **Number of products/services affected:** Will it enhance cross-functional focus?

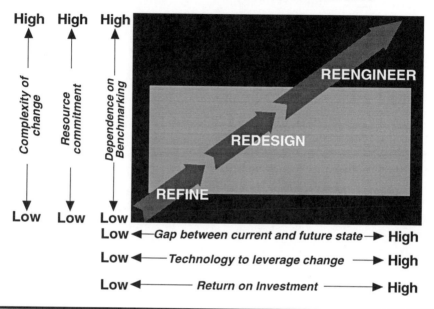

FIGURE 3.5

Having answered these questions, the next thing the QLT must determine is to what extent a process needs improvement. In other words, to what degree does the process need change in order to best serve customers and achieve the company's strategic objectives? Typically, process improvement initiatives fall into three broad categories (Figure 3.5):

▪ **Refine:** Through incremental improvement
▪ **Redesign:** Through changes in the order or way in which existing activities are performed
▪ **Reengineer:** Through assessing what has to be done and what is the best way to do it

The difference among the three tends to be the scope or breadth of the process and the level of effort required. In refinement through incremental improvement, the goal is to attain a new level of performance which exceeds previous levels by improving what went before—to refine, but not radically change, the existing process. This refinement is all well and good **if** the resulting process is capable of producing the desired results.

Refinement initiatives typically take only a short time to complete in the improvement process and usually produce positive change. However, the true litmus test is always results. If the process is not capable of being modified in order to help achieve strategic goals, simply refining it is not enough.

The second level of process improvement is **redesign,** i.e., the modification, fixing, or reordering of an existing process. As mentioned earlier in Chapter 2 and repeated below, redesign projects involve greater and more complex change, the span of process improvement is larger, and benchmarking is involved to a great extent. Impact on the organizational strategy and mission is greater, and the redesign approach generally requires increased support from information systems and technology.

"**A** mistake that some organizations make is implementing expensive information technology that merely automates existing practices. They don't change or improve the business process; instead, they simply spend a lot of money to automate it."

Michael Hammer and
James Champy[1]

If not watched and managed well, redesign can absorb significant resources, yet still fail. Simply put, redesign involves major change to the existing process in order for it to achieve its desired results. However, as currently implemented, redesign all too often is a cautious adjustment of limited scope. Such projects often fail. A truly great organization is never satisfied with its current performance or with minor modifications unless its level of performance is already world class.

"**T**he slow cautious process of incremental improvement leaves many organizations unprepared to compete in today's rapidly changing marketplace. But reengineering helps organizations make noticeable changes—'quantum leaps'—in the pace and quality of their response to customer needs."

Robert Janson[2]
"How reengineering transforms organizations to satisfy customers"
National Productivity Review

Too often, organizations that implement process improvement programs initially focus only on process refinement or redesign without regard to the scope of change needed. This can be a mistake. For example, after an analysis of an organization's business processes, it may be determined that one or more processes will never be capable of meeting the demands that are placed on them. Thus, the approach required is to fundamentally change the way an organization does business.

That improvement approach is called **process reengineering.** It aligns operational plans, business processes, people, and information technology to support a company's strategic vision. Process reengineering puts people, processes, and technology on the same team and focuses them on a common goal.

Process reengineering involves completely rethinking the operation and supporting information systems. Its goal is to produce dramatic improvements by identifying, benchmarking, and improving key business processes. Given the three levels of process improvement, the QLT must consider its strategy, its customers, and its competitors and decide to what extent the targeted processes must be changed. It then charters the teams to do it.

For each project, the QLT identifies process boundaries, key strategies (refine, redesign, or reengineer), process owners (people with the authority to change the process, if necessary), process measures, preliminary target levels to be achieved, and potential team members in a written charter for each team. The charters, such as that shown in Figure 3.6, are the core of the process improvement plan and serve as the guiding documents for process improvement teams during the deployment phase.

> "**M**ost change efforts start with what exists and fit it up. Reengineering, adherents emphasize, is not tweaking old procedures and certainly not plain-vanilla downsizing. Nor is it a program for bottom-up continuous improvement. Reengineering starts from the future and works back, as if unconstrained by existing methods, people, or departments."
>
> Thomas Stewart[3]
> *Reengineering: The Hot New Tool in Managing*

The final step in creating a process improvement plan is drafting an implementation plan that captures all the information, decisions, and policies that have been agreed upon by the QLT during the *prepare* and *plan* phases. The QLT will use this plan to communicate its intentions to all employees.

In conclusion, the objective of the process improvement plan is to improve those processes that are important to the customers and the objectives of the company. Focused Quality Management is what it takes for quality initiatives to get positive results. Work gets done (or does not get done) and organizations succeed (or fail) based on what

Sample Team Charter

DATE: _____

Team Leader: _____

Team Facilitator: _____

Team Members: _____

_____

_____

_____

Process Owner: _____

Goal:

Decreased patient waiting time in outpatient areas.

Boundaries:

- Implementation recommendations can be applied to all outpatient areas (unless clinic specific.)
- Recommendations can be accomplished with existing personnel.
- Recommendations will be based on TQM philosophies and analyzed according to a data driven approach.
- Recommendations will be in compliance with standards.
- The team will seek input from all services involved in the implementation of recommendations.

Strategies:

- Assess/review current medical center policies and procedures related to outpatient appointment, triage, waiting, MD visit, dismissal, etc.
- Identify all outpatient areas, and those processes that are uniform throughout each clinic.
- Determine the resources currently available to deliver patient care, both direct and indirect, in the outpatient area.
- Determine the clinic, and/or process which has the largest impact on patient waiting time.

Outcomes:

- All patients will be seen within 30 minutes of scheduled appointment time.
- Patient satisfaction will increase as determined by measurement tool.
- Provider satisfaction will increase as determined by measurement tool.
- 100% of medical records will be complete and available for scheduled outpatients at their appointment time.

Preliminary Time Frame:

Report to Quality Leadership Team within 120 days.

FIGURE 3.6

occurs within business processes. By linking strategic goals, quality programs, and the business processes that most affect customers, FQM does work, and organizations get results that make a difference when process improvement teams carry out the *deploy* phase, which is described in the next chapter.

In addition to developing a process improvement plan, the leadership also prepares an organizational improvement plan. This plan is aimed at making the necessary structural and procedural changes needed to support the FQM initiative. It addresses the training, communication, reward and recognition, and measurement systems. These systems are what is needed to enable the organization to perform at a higher level of quality. Specific recommendations that have proved to be effective components of organizational improvement plans will be discussed in Chapter 5.

BIBLIOGRAPHY

1. Hammer, Michael and James Champy, *Reengineering the Corporation: A Manifesto for Business Revolution,* New York: Harper, 1993.
2. Janson, Robert, "How Reengineering Transforms Organizations to Satisfy Customers," *National Productivity Review,* p. 52, Winter 1992/93.
3. Stewart, Thomas A., "Reengineering: The Hot New Managing Tool," *Fortune,* p. 41, Aug. 23, 1993.

4

DEPLOY: MAKING PROCESS IMPROVEMENTS HAPPEN

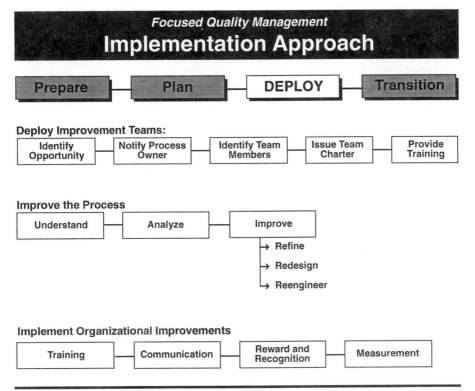

Focused Quality Management
Implementation Approach

Prepare — Plan — DEPLOY — Transition

Deploy Improvement Teams:

| Identify Opportunity | Notify Process Owner | Identify Team Members | Issue Team Charter | Provide Training |

Improve the Process

| Understand | Analyze | Improve |

→ Refine
→ Redesign
→ Reengineer

Implement Organizational Improvements

| Training | Communication | Reward and Recognition | Measurement |

FIGURE 4.1

The *deploy* stage, the third stage of Focused Quality Management (FQM), as shown in Figure 4.1, puts the strategic quality plan into action and empowers process improvement teams by providing clear objectives, in focused charters. It builds the organization's capability to improve itself to meet customer expectations and makes it possible to measure progress and determine success or the need for further improvement. In this step, team activities to improve the previously identified key business processes occur. Process improvement teams are formally chartered and trained. The teams' goals, as specified in their charters, may be to improve, through refinement or redesign, selected processes to reduce errors, delays, defects, and costs while increasing customer satisfaction. In the case of reengineering teams, quality tools are used to develop the most cost-effective method to meet customer expectations.

In general, the improvement team studies the process to understand how it works and how it needs to be improved. The team then analyzes process performance in order to identify recommendations for how to improve the process. In keeping with the teachings of Juran, Deming, and others, these changes are actually implemented, and their new performance is measured to determine the impact of the change. When the results are positive, all this effort culminates in a celebration of the team's accomplishments.

The benefit of the above-mentioned steps and the preparation and planning preceding deployment in the FQM process is *focusing*. Teams are focused on the processes that need improvement. Unlike some other approaches to quality, teams are provided direction about what they are to accomplish. For example, proponents of quality circles often advise upper management to give very little direction to quality circle participants, the theory being that direc-

"**W**e are convinced that every company faces specific performance challenges for which teams are the most practical and powerful vehicle at top management's disposal. The critical role for senior managers, therefore, is to worry about company performance and the kinds of teams that can deliver it. This means that top management must recognize a team's unique potential to deliver results, deploy teams strategically when they are the best tool for the job, and foster the basic discipline of teams that will make them effective. By doing so, top management creates the kind of environment that enables team as well as individual and organization performance."

Jon R. Katzenbach and
Douglas K. Smith[1]

tion will stifle creativity. The results can be frustrating to both the team and management. The following dialogue exemplifies the most ludicrous of such an unguided approach to quality circles:

> When members of a quality circle ask, "What do you want us to work on?" upper management responds, "What do you want to work on?" When the quality circle counters, "Well, what do you think is important?" upper management is prompted to answer with, "What do *you* think is important?" When the quality circle says, "Why do you always answer a question with a question?" upper management responds with, "What do you mean by that?"

Not surprisingly, in the quality circle scenario described, upper management ends up asking why the teams do not achieve meaningful results. The problem is not that the team failed; rather, management failed because the team lacked the direction they needed from management to help them succeed. Most employees want to know what is expected of them and what is important to the organization so they can focus their work toward that end. With *focused* quality management, that guidance is provided and strategic results are the outcome.

Why teams? What is their value? Under what circumstances can teams make a positive difference? What can the Quality Leadership Team (QLT) do to enhance their effectiveness?

Teams properly commissioned and supported have proven time and again that they have the ability to deliver tremendous organizational results. Process improvement teams comprised of a cross-section of employees are the only way to improve cross-functional processes. As stated throughout this book, improving cross-functional processes is how organizations achieve their strategic goals and objectives.

At Schlumberger Well Services, North America, Larry Gutman, V.P. North America Operations, and Mike Robson, V.P. Quality, involved themselves and the entire senior and mid-level management staff in their improvement process. Not only did the management team develop the vision, mission, and values and identify critical success factors, but they also attended quality team training and participated as members of process improvement teams. The results of this level of leadership commitment are impressive. Management buy-in is strong, as shown by support for deploying the process throughout the organization. In addition, the company, which has always encouraged and rewarded individual behavior, is shifting its culture to focus on cross-functional, team-based solutions.[2]

CHARTER THE IMPROVEMENT TEAMS

During the *plan* phase the QLT reviewed the feedback from the organizational assessment and prioritized what processes needed improving first. Based on these priorities, the QLT also created criteria for what kind of cross-functional project teams and which approach (refine, redesign, reengineering) were needed and drafted team charters. Those charters gave clear direction to the teams, including process boundaries, decision-making authority, time frames, strategies to define the process, and desired outcomes. The specifics to accomplish the task are left to the team, but the responsibility remains with management. Accordingly, a member of the executive committee/quality council needs to take ownership for coaching and support.

In order to identify an effective approach to the task at hand, the process improvement team needs to give thought to its relationship with the QLT. The following are examples of some questions that both should discuss and agree upon:

> "**M**ost successful teams shape their purpose in response to a demand or opportunity put in their path, usually by higher management. This helps teams get started by broadly framing the company's performance expectation. Management is responsible for clarifying the charter, rationale, and performance challenge for the team, but management must also leave enough flexibility for the team to develop commitment around its own spin on that purpose, set of specific goals, timing, and approach."
>
> Jon R. Katzenbach and
> Douglas K. Smith[1]

- How and how often will the process improvement teams communicate with the QLT?
- What steps do the process improvement teams need to go through to assure they understand their charters?
- How will the process improvement team make decisions?
- Who does the process improvement team report to and how will it get assistance from others?
- Who should be notified if team deadlines cannot be met?

As to the composition of the process improvement team, there needs to be a combination of team leaders, facilitators, and members. Obviously, different skills are required of each of these groups. Team leaders are responsible for accomplishment of tasks and facilitators are responsible for an effective group process. Team leaders need to be knowledgeable about the process being improved, although they should not be too vested in the status quo and/or uncomfortable with change. Leaders also have responsibility for working closely with the process owner (the manager responsible for maintaining the process flow), because in most cases, changes cannot occur without the support of this individual.

Facilitators, on the other hand, are not required to be as knowledgeable in the specific operation of the process the team will be working to improve. In fact, they often are neutral in the activities and usually do not voice their opinions, but simply guide the discussion and decision-making process.

Most team members are people with knowledge of the existing process, but creativity can be enhanced and the ultimate outcome improved if a member is added who has knowledge of or experience in similar processes handled in a different fashion. A member with quite different experience may even be added to further stimulate the creative process.

The following are some criteria the QLT should consider in selecting team members:

- Knowledge of the process
- Broad representation by affected groups
- Variety of levels represented
- Proportion of management to professional to front-line personnel
- Size of team
- Communication skills
- Problem-solving skills
- Interpersonal skills
- Managerial skills
- Leadership skills
- Vested vs. no vested interest
- Creativity
- Ability to be objective

Once team members have been selected and charters issued, plans must be made for training in quality awareness, effectiveness of meet-

Sample Mission Statement

Our mission is to identify opportunities to reduce patient waiting time in outpatient areas so that:

☑ All patients are seen within 30 minutes of their scheduled appointment time

☑ Patient satisfaction is measured and improved

☑ Provider satisfaction is measured and improved

☑ Medical records availability and accuracy are at 100%

FIGURE 4.2

ings, and quality and process design tools. Those selected to be team leaders or facilitators should also receive training in facilitating meetings, team dynamics, and managing group processes.

Having completed training, one of the first tasks for the process improvement team is to create its **mission statement.** A sample statement is shown in Figure 4.2. The objective is to develop the team's understanding of its purpose and allow the members, in their own words, to communicate it to the QLT and the process owner. This mission statement is derived from the team charter and stated in the team's language.

Thus, it provides the opportunity for clarification or redefinition by the QLT. It also helps the process improvement team, because the mission statement guides them in preparing an action plan to direct and control their activities throughout the improvement initiative.

Another benefit of creating a mission statement is the clarity and commitment gained by the process improvement team as to its purpose. Without a clearly stated purpose and common commitment, team members tend to act independently rather than rallying

A cross-functional team looked at disposable linen use in a large hospital. The team discovered that they could reduce hazardous waste, increase safety, and provide satisfaction while giving increased comfort to patients and saving $38,000 per year. They also discovered that the cross-functional team made implementation of their solutions much easier.

Sample "Rules of Trust"

- ☑ **All meetings will begin and end on time.**
- ☑ **Team members will attend all meetings and be on time.**
- ☑ **We will listen courteously to other team members ideas.**
- ☑ **Everyone will share in the duties of the recorder.**
- ☑ **Agendas will be provided 48 hours before all meetings.**
- ☑ **Criticize ideas, not individuals.**
- ☑ **Assignments will be completed on time.**

FIGURE 4.3

together around a common goal they believe in and wholeheartedly support.

Another task facing the process improvement team at this early stage is the development of its **rules of trust,** or ground rules for how members will conduct themselves (Figure 4.3). Research on effective team performance shows that clearly stated rules of trust are key elements of all successful teams. To create their rules of trust, the process improvement team members discuss and come to an agreement about how their meetings will be run, how their members will interact, and how their behavior will be judged acceptable. This kind of up-front discussion is invaluable in creating a sense of mutual accountability for the success of the effort.

A final step to be taken by the process improvement team is the establishment of a preliminary **work plan.** The work plan consists of a list of activities and times required

> "**T**he essence of a team is common commitment. Without it, groups perform as individuals; with it, they become powerful units of collective performance. This kind of commitment requires a purpose in which team members can believe. Whether the purpose is to 'transform the contributions of suppliers into the satisfaction of customers,' to 'make our company one we can be proud of again,' or to 'prove that all children can learn,' credible team purposes have an element relating to winning, being first, revolutionizing, or being on the cutting edge."
>
> Jon R. Katzenbach and
> Douglas K. Smith[1]

to accomplish the team's mission. It reflects a logical approach to understanding the process, analyzing the process, testing, and, if successful, implementing the process improvements. As shown in Figure 4.4, the work plan is also a very effective management tool. It provides ongoing direction and control of the team's activities by establishing dates for review of the team's interim progress by the QLT.

> "**A**t its core, team accountability is about the sincere promises we make to ourselves and others, promises that underpin two critical aspects of effective teams: commitment and trust."
>
> Jon R. Katzenbach and Douglas K. Smith[1]

Because it is prepared by the team, the work plan provides a reality check on the time frame originally established by the QLT in the team charter. Preparing the plan requires the team to give thought to the time it will take to properly understand, analyze, and make improvements to the process it is evaluating. Planning interim review dates provides the team with specific checkpoints to review progress with the process owner or representatives from the QLT.

In closing, the following are key to the success of a process improvement team:

- A well-defined charter provided by the QLT
- Team participation in the development of a team mission statement, rules of trust, and a work plan
- Development and maintenance of a productive team environment
- Understanding and use of the quality tools and the concepts of process improvement, i.e., refinement, redesign, and reengineering
- Troubleshooting skills, feedback mechanisms, and information sources

IMPROVE THE PROCESS

Once the teams have been chartered and trained, it is time to move to the heart and soul of the *deploy* step, which is to focus the team on actually improving the process. As stated earlier, teams are typically asked to reduce errors, delays, defects, or customer complaints.

To make these improvements, the team begins by evaluating the process-specific assessment data in the context of customer expecta-

TQM Action Plan For _____ Project

Action	Responsible person and/or Function	1	2	3	4	5	6	7	8	9	10
PHASE 1: UNDERSTAND THE SITUATION											
Identify Key Process		▓									
Identify Customers											
Identify Expectations											
Identify Variation			▓								
Diagram Process Flow				▓							
Set Priorities/Goals/Objectives				▓							
Cause & Effect/Pareto					▓	▓					
Measure Process							▓	▓			
PHASE 2: ANALYZE											
Construct Histogram(s)							▓				
Develop Run Chart(s)							▓	▓			
Construct Scatter Diagram									▓		
Develop Control Chart(s)									▓	▓	
PHASE 3: IMPROVE											
Propose Improvements/Plan								▓			
Select Improvements/Plan								▓			
Implement Plan									▓	▓	
Collect and Analyze Data									▓	▓	
Compare Results										▓	
Document/Standardize											▓
Project Review/Approval			▓			▓		▓		▓	▓

Note: Schedule shown is for illustration only. Project activities and time lines will vary depending on the process and scope.

FIGURE 4.4

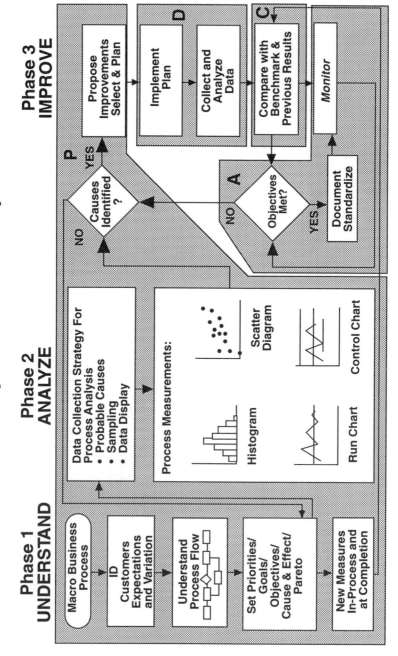

Quality Team Roadmap

Phase 1 UNDERSTAND · **Phase 2 ANALYZE** · **Phase 3 IMPROVE**

FIGURE 4.5

tions and the "as is" environment. The team then follows a systematic approach for analyzing and improving the process to reach the "to be" goal. This forces the team to rely on the use of data so the members can understand, analyze, and improve their process for lasting results, as illustrated in Figure 4.5.

In other words, a team evaluates a process to understand just what the process is supposed to be doing. It then spends time analyzing by asking "What does the data that we gathered say about how the process is actually working?" Once a team understands how the process works and has measured its current performance, the next step is to decide how it can be improved and then test the improved process, whether it was refined, redesigned, or reengineered. Based upon these results, the team presents recommendations for approval and implementation.

The key to improving processes is asking the right questions and using simple analysis tools that enable teams to understand how the process works currently, to analyze what is causing the process to function inefficiently or ineffectively, and to offer recommendations that will truly improve the process.

> "**A**sking questions and looking for answers without fear is how true solutions are found."
>
> Lyell Jennings

Teams should ask the following series of questions when attempting to understand, analyze, and improve a process:

1. Who are the customers and what are their expectations?
2. What is a valid measure of how well the process works?
3. How does the process work currently?
4. What things in the process are causing problems?
5. What changes are needed in the process to satisfy its customer?
6. How can each team member begin to understand the process? What tools and techniques can he or she use to answer such questions as:

   ▪ Are there any predictable or demonstrable patterns or trends about the process that might help me see where/when bottlenecks occur?

   ▪ How stable is the process? How often does it vary and why? How much variance is acceptable? At what level of variance should I become concerned?

- Is there a correlation between events in the process? Is there a correlation between time of day and error rate? Do more errors occur between midnight and 6:00 a.m.?

- How long does the process take to complete (cycle time)? What is the average time? What is the variation, i.e., shortest and longest intervals?

- Which problems are causing the most complaints from the customers?

- What is the process cost profile?

7. What kind of process improvement is necessary? Did the QLT determine if:

- The process is basically working well and simply requires monitoring and fine-tuning to make incremental improvements

- The process has some redeeming qualities that should be kept, but still requires some major redesign

- The process is in such poor condition that it needs to be reengineered

Teams can best understand the process they have been chartered to improve by asking the following question: "Who are the customers and what do they want?" It is surprising how many times people, especially those who are part of a process, have never asked this particular question. Yet it is a key question. Once the team has asked the customers what they want, it usually has a pretty good idea of some specific problem areas, because the customer will also tell the team what he or she is unhappy about.

Before something can be fixed, however, one must know something about how it currently works. The following example describes a restaurant where customers had complained about the food service.

A major hotel wanted to improve its restaurant's food service. They wanted to start with a simple process. Because of numerous complaints about the chicken parmesan, they decided to look at the process by which this particular entree was prepared.

They found that the chicken parmesan moved about 160 feet from the loading dock to the dining room where it was served. It arrived at the loading dock as a chicken breast and ended up on the dining room table as chicken parmesan. That's the good news.

The bad news is that when they looked at the process by which the chicken parmesan was made, they found out that it moved four floors vertically through the hotel, only to return to the same level where it had started. It was handled by about seven different people, moved through numerous temperature zones, had numerous subprocesses (such as the cheese subassembly and the sauce subassembly process), and probably over 250 steps when you got down to the nitty-gritty. But the fascinating thing was that in physical distance traveled in order to go from the loading dock to the dining room table (160 ft.), the chicken parmesan in reality traveled 1/4 mile.

When the team recognized the complexity of the process used to make the chicken parmesan, they realized that it wasn't surprising that they occasionally made a bad meal, but it was *amazing* that they ever made a good one.

This story illustrates that processes used to provide products and services to customers are usually much more complex than ever imagined. More people are involved in the process than most people would have thought possible. Also, processes typically have numerous instances of redundancies and non-value-added activities. It is therefore not surprising that many people conclude, after analyzing a process, that it is amazing that it is ever done right and not surprising that it is sometimes done wrong.

As demonstrated by the chicken parmesan story, the best way to answer the question "How does the process work?" is through the use of a **flow chart** (Figure 4.6). It is not the only tool, but it is one that facilitates consistent communication. Soon everyone knows the flow of activities and sees that management places importance on understanding how processes work within the organization.

A team from the Dallas Veterans Administration Medical Center flow charted the patient admissions process, modified the process, and eliminated unnecessary procedures. The result was a decrease in admission wait time from over two hours to twenty minutes or less.

A good way to put together a flow diagram is to begin by brainstorming. People associated with the process get together to identify all of the things that must happen to create a product or service. Their ideas are recorded on individual slips of paper and displayed on a wall or table. Then the group moves the steps around until everyone feels

Flow Chart Example:
Lab Work Order Process

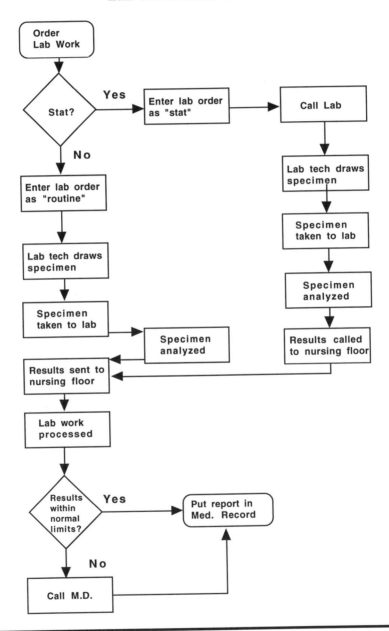

FIGURE 4.6

Cause and Effect Diagram

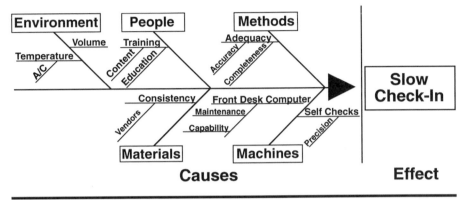

FIGURE 4.7

that an accurate picture of all the steps in the process has been captured.

Once the team understands how the process works, it will have a better idea of what is working well and what is not and will be able to set goals for how the process should be changed in order to satisfy customers. For example, a goal for the food service problem discussed earlier would be, "We will shorten our food preparation process and eliminate these specific unnecessary steps so we can serve our customers in less than ten minutes."

At this point, it is crucial to get a lot of information on the table about what is causing the process problems. One technique for doing this is brainstorming. Each team member is asked to record all the causes that could be related to this problem on slips of paper. Once this is done, the group members use the slips of paper to organize their ideas into similar categories

A Phoenix Medical Center used cause-and-effect diagrams to identify root causes for post-operative infection rates that exceeded the national average of 2%. They then developed solutions that lowered the rate to 0.4%, less than one fourth of its previous level.

Once the causes have been organized into groups that relate to each other, it is easy to create a cause-and-effect or fishbone diagram, like the one in Figure 4.7. The **cause-and-effect diagram** enables the group to analyze the process for dispersions or breakdowns. Its purpose, obviously, is to relate causes and effects. "A cause-and-effect diagram is a powerful tool. It can help resolve problems because it directs the treatment toward root causes, not symptoms."[3]

Now that the process as well as the causes of the problems are understood, the next step is to analyze the process so that the team can verify its assumptions and measure how the process currently works. Good questions to ask include: "What is the error rate?" "What is the cycle time for this entire process?" "What is the cycle time for this particular step in the process?" "How often does the cause impact the outcomes?" The goal is to look for clues to identify where time or resources are being wasted or error rates are too high and then determine if the causes identified are really at fault.

This evaluation leads to asking: "What is the measure of how well the process works?" For every output of a process, there should be a generally agreed-upon way to measure it. For example, there may be a problem getting customers to pay their bills. Once a flow chart of the billing process has been completed and all the steps from the input (accurate order entry) to the output (a check from the customer) have been described, then how the process is working can be measured.

One measure could be the percent of receivables more than 90 days past due. However, that may be an effect rather than a cause. The next question is, "Are the customers not paying their bills because they are deadbeats, or do they have a problem with their bills?" The quality of the bill (i.e., accuracy) could be the true problem, so there should be some measure of that particular characteristic.

> **A** team used data analysis to identify the highest impact causes for surgery cancellation. They developed ways to reduce cancellations by 17% and saved $350,000.

Assuming that some valid measure has been identified, it is now time to conduct an in-depth analysis of the process. This analysis will enable an understanding of the process by using simple statistical tools and analysis techniques to answer questions raised earlier, such as:

- Are there any predictable or demonstrable patterns or trends about the process that might help me see where/when bottlenecks occur?

- How stable is the process? How often does it vary and why? How much variance is acceptable? At what level of variance should I become concerned?

- Is there a correlation between events in the process? Is there a correlation between time of day and error rate? Do more errors occur between midnight and 6:00 a.m.?

- How long does the process take to complete? What is the average time? What is the variation (i.e., shortest and longest intervals)?

- Which problems are causing the most complaints from the customers?

The first thing to do is create a **histogram,** which is a graphic summary of how the process works and varies in its performance. A histogram pictorially displays such things as how long it takes on the average to complete the process, the shortest time the process ever took, and the longest time it ever took.

The value of a histogram is that it allows a quick identification of how the process performs day to day; thus, its present capability to meet customer requirements is known. For example, the data table shown in Figure 4.8 provides little insight and is not very user friendly. However, when displayed in a histogram (at the right in the figure), the process performance level and variation are very clear.

Because all organizations have limited resources, they need to get the most out of their process improvement activities. To do this, the next question is asked: "Which of these causes has the highest impact on the organization?" Impact in this case means what makes the most customers dissatisfied or what is the biggest cause of waste.

What are the most frequently occurring/recurring, defects, etc.? The tool for answering this question is the **Pareto chart.** Pareto charts show that the most important effects are the result of very few causes.

"It is well known that 80% of the funds contributed to charity come from only 20% of the possible sources."

John Burr[4]

They also are a graphic way of showing which of a number of process problems occurs most frequently and has the greatest impact. Thus, Pareto charts clearly show which types of causes are the reason for the majority of problems in the process. For example, a team used Pareto charts to help it understand the causes of customer complaints and how frequently they occur (Figure 4.9). Armed with this information, the process improvement team was able to prioritize its efforts and work on the problems

Histogram

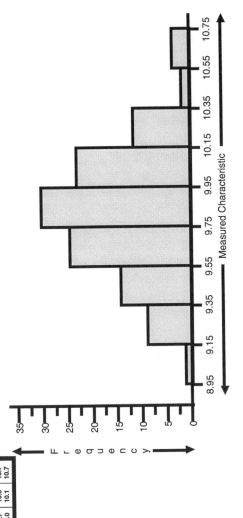

FIGURE 4.8

Pareto

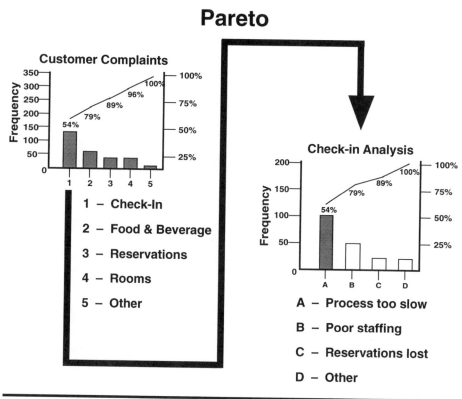

FIGURE 4.9

that most negatively impacted the customers; in this case, two Pareto charts were used to zero in on the root cause.

Pareto charts do not just happen. In order to prepare them, data must first be gathered so that they can be summarized in a chart. The data collection tool for this purpose is usually a **checksheet.** Placing a checksheet in the areas of the organization where the process is performed will help quantify the relationship between the causes and the effects of the process. Asking people to check each time one of the causes occurs is easy and starts the collection of objective data. The data from the checksheet are tallied and drawn on a Pareto chart. The team can now focus on the cause with the highest impact in order to eliminate the root cause of the problem.

Another valuable tool that improvement teams can use is the **scatter diagram.** Scatter diagrams are used to determine the relationship

Scatter Diagram

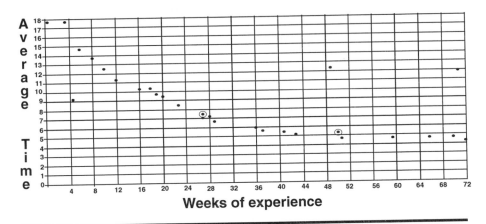

FIGURE 4.10

between two variables. A scatter diagram may be used to help determine whether a cause-and-effect relationship usually exists between two variables. However, it is important to note that it will only indicate whether or not a relationship exists between two variables; it will not indicate whether one is the cause of another.

The scatter diagram in Figure 4.10 was used by a process improvement team for a hotel to determine if a correlation existed between the average weeks of experience of the front desk staff and the speed of checking-in hotel guests. From the diagram it appears that a strong correlation exists between experience and speed. As new employees gain skills, their speed increases and check-in time drops. From the graph in Figure 4.10, it seems that once an employee has 36 weeks of experience, further improvement is very slight.

The last two quality tools addressed here are run charts and control charts. Both show events in time order. A run chart is the basis of a control chart.

The **run chart** is a simple way to show graphically the variation in process output over time. It is a good tool to illustrate trends and results. Time is on the x axis and frequency or some other measure is on the y axis.

The run chart in Figure 4.11 shows the number of customer inquiries per 100 invoices on a weekly basis. It appears that something happened

Run Chart Example
Customer Inquiries

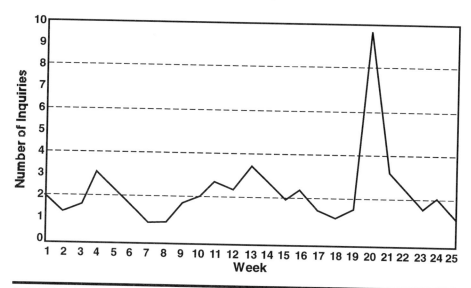

FIGURE 4.11

during week 20. Upon investigation, the team found that in week 20 the firm changed billing cycles without informing its customers. The result of that decision is clear from the chart.

With **control charts,** the data are further evaluated to determine the natural limits of the performance of the process and to represent those limits on the chart (see Figure 4.12). As noted earlier, a process will vary over time. A control chart shows the expected range of variation. Comparing that with a performance standard can help the team determine an acceptable level of variation and identify when there really is a need for concern. It also allows the team to interpret the cause of the variation. A control chart helps the team to determine whether or not a process is stable and predictable and allows them to identify when variation is a result of common causes. It also helps the team to clarify when the worker needs to react to an unusual or special cause.

In essence, a control chart is a tool for showing trends and the location of upper and lower control limits that reflect the normal bounds of performance. This is important because if a process is

Control Chart

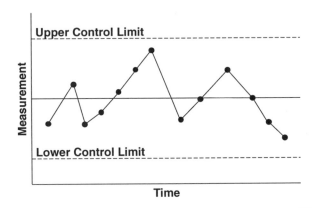

FIGURE 4.12

operating within normal limits and meeting customer needs, it may not be wise to spend time and energy to change it; instead, it may be better to just let it run as is. However, if something is outside of normal limits or the normal performance of the process is inadequate, then the team would probably want to take action and fix it, or would want

> **W**hen improving a process, 95% of process problems can be solved using the seven simple tools of quality.

the people down at the process level to be aware of the abnormal performance so that they know when they have to step in and take action.

TEST THE IMPROVEMENTS

When the team is finished with analysis, it is ready to take the most important step in the *deploy* phase: implementing changes to improve the process. Because it has listened to the voice of the customer and to the voice of the process, the team is now ready to act.

The team begins by asking: "Have all the causes of the problems in the process been identified?" If the answer is yes, the team then generates ideas/solutions for improving the process.

After entertaining numerous possible solutions, the team selects what it believes is the best one and develops an implementation plan for testing the proposed solution by implementing the change on a trial basis.

At this point the team has an extensive understanding of the current process and what needs to be done. It is in a position to recommend to the QLT the level of improvement that will be needed to ensure that the process achieves maximum results. Much has been said about the current focus on process reengineering, with its concurrent reductions in the work force and promises of quantum leaps in productivity and earnings. This approach does not presuppose that reengineering or anything similar is the answer. Rather, it assumes that after an in-depth assessment is done, management and the process improvement team together can determine what is needed.

If a process is failing to perform as desired, management has three basic choices, as described earlier: to **refine** (make incremental improvements to an existing process), **redesign** (change the shape, timing, or relationship of existing process segments), or **reengineer** (start with a blank piece of paper and create a process that achieves the desired outcomes, regardless of what used to be).

Refinement is a common approach used by many teams, because most processes simply need a more efficient or effective way of accomplishing each process step. Redesign is often employed when the process is accomplishing tasks in serial, consecutive fashion, but parallel processing and simultaneous activities may make more sense and save time and money.

Reengineering starts afresh. First, the process outcome is determined. Then the best (cheapest, fastest, most reliable, etc.) way to accomplish it is examined. As a fresh approach, reengineering often yields tremendous increases in performance simply because it breaks the mold of conventional thinking and opens new avenues for improvement.

Reengineering also has a price. Typically, firms committed to reengineering spend hundreds of thousands of dollars conducting assessments and process reengineering studies. Fortunately, the payoff is often greater, sometimes as high as a 10-to-1 return when the new process is put in place. Redesign yields similar results, although of lesser magnitude (4 to 1) in dollar payoff. Least in order of magnitude, but still profitable (2 to 1), are the results of process refinement.

It may seem tempting to reengineer (a 10-to-1 payoff) and not waste time on refinement (a 2-to-1 payoff). The real driver of which approach to take, however, is the outcome of the assessment. The assessment is

Adolfo Guerrero Camacho of the State Commission of Public Services of Tijuana (CESPT) responded to the direction of Lic. Jose Osuna Millan, Director General of CESPT, to develop a quality improvement process by ensuring that

■ All senior managers received training in the philosophy of total quality

■ Workers received training in quality philosophy and the Japanese-developed 6S method:

Seiri	To fix the process, eliminate unnecessary activities, correct discrepancies, and follow procedures
Setton	To put things in proper order, clearly marked
Seisow	To make facilities and work areas clean and orderly
Seiketsw	To keep them that way
Shwkam	To make the preceding a way of life
Shitsuke	To maintain the discipline to do the above

Because workers were encouraged to put the 6S approach in place, the results were immediate and impressive. Site visits revealed clean, orderly workman-like facilities and people. With that foundation of enthusiasm in place, management solicited employee ideas for improvement.

designed to discover, based on an analysis and understanding of the existing process, what is needed. If a process is basically sound, refine it. If a process can be substantially improved, redesign it. If a process is fundamentally flawed, reengineer it. In all cases, the judgment should be made *after* the appraisal, not before. Let the process guide the actions.

The process improvement team should make a recommendation after completing its process analysis as to the scope of the process improvement effort. The QLT should review those recommendations and select the best course of action.

As stated earlier, the greatest potential payoff comes from the reengineering approach; however, as shown in Figure 4.13, that approach also requires the greatest degree of alignment with the overall strategic plan, broadens the scope of the change in the process, demands the most information systems-based and often technological interface, and relies most heavily on benchmarking other similar processes. Not surprisingly, it requires the greatest commitment of resources because, as most leaders understand, there is no free lunch in the pursuit of excellence. Balancing all this is the fact that along with the greatest resource demand comes the highest potential payoff.

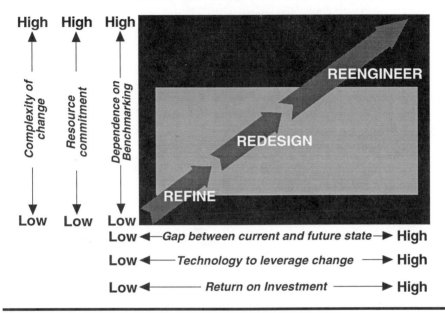

Deploy Step
Three Tracks for Measurable Results

FIGURE 4.13

Refinement of existing processes requires less of a resource commitment. However, the process must be fundamentally sound in order to justify this approach. Simply put, refinement of a bad process through incremental improvements allows a company to do the wrong things well. If the process is sound, incremental improvements will let a company do the right things in the most effective and efficient manner.

Redesign of an existing process goes beyond the incremental approach of refinement by searching for better ways to do the same things currently being done. Can the total time to accomplish an activity be shortened by doing some things in parallel that were previously accomplished sequentially? Could cycle time be shortened or the quality of the process increased by changing when and how particular steps are done? This approach goes beyond incremental improvements, but stops short of total reengineering of the activity. The resource commitment is greater than refinement, but less than reengineering, and the potential payoff varies accordingly.

The choice is up to the particular organization, but that choice should be driven by the results of the assessment. If the process is basically sound, refine it. If the process is weak, redesign it. If the process is broken (or one does not exist), reengineer it.

Fortunately, serendipity will sometimes determine which approach should be used. For example, a hotel chartered a team to improve the check-in process by reducing the time required to check in (a factor that their most valued frequent-traveler customers said was very important to them). The team looked at better training of front desk personnel, faster computers, and better printers (i.e., the refinement approach) to gain small improvements in speed.

The results were good, but led the team to consider getting even greater improvement by setting up a special procedure (redesign) to handle members of the hotel's frequent-traveler club.

The breakthrough came when a member of the bell staff asked, "Why do we check them in at all?" The response was a vehement, "Because." He persisted. "Look, we let them leave without any hassle with our express check-out service when they owe us money. So why do we make it tough to check in to this place?" He went on to say, "We know they're coming, they have a reservation. We know who they are, what credit card they are using, and what kind of room they want. Why can't I just hand them their room key when I pick them up at the airport?" The rest of the team was speechless. A non-front desk person had seen the real goal: a traveler in the right room quickly. Because he did not feel tied to the existing process, he simply asked, "What is the smartest way to do this?"

That question is the very essence of process improvement. The job of a leader in the organization is to ask that question every day and in every way. The answers may be surprising, but the results are worth the effort.

During the implementation and after, the solution is measured using the exact same measurement tools used earlier in order to accurately compare before and after results. With this approach the team has tangible results available in order to determine whether or not the objectives were met. In essence, team members are trying to find out if their trial solution worked. Did it improve the process? If so, by how much?

If the answer is yes, the team documents the process changes, reports them to management, and gets approval to standardize how the process is going to work from then on. Standardization and monitoring are implemented to ensure that the solution continues to make a positive improvement.

On the other hand, if the trial solution does not work, the team tries to determine why the solution did not produce the expected results. Team members question whether or not they identified the actual root cause of the process problem, or if they simply picked the wrong solution. Making this determination requires analysis of the data and may lead the team to use the cause-and-effect analysis again. If, in fact, the team finds that it did identify the root cause but just came up with the wrong solution, it reevaluates the solutions and tries again.

A final and very important part of the *deploy* phase is celebration. After the process improvement team has successfully improved the process, it is crucial that the QLT celebrates and recognizes the efforts of the improvement team. Celebrating demonstrates a sincere appreciation for the team's contribution to the organization. It also demonstrates management's commitment to pursuing quality as a means of achieving the organization's goals, as well as a commitment to the collaboration required to improve how the organization conducts its business.

In conclusion, the process improvement efforts in the *deploy* stage are where the "rubber meets the road." FQM is no longer something talked about at the highest levels of the organization, but is now a reality throughout. Teams have been given a very important task and they have delivered. They can be proud of their accomplishments, not only because they have been personally recognized, but also because they have tangible measurements of improved customer satisfaction and profitability.

BIBLIOGRAPHY

1. Katzenbach, Jon R. and Douglas K. Smith, "Discipline of Teams," *Harvard Business Review,* pp. 111–120, March–April 1993.
2. Gutman, Larry and Mike Robson, personal communications, Spring 1994.
3. Carter, Carla, "Seven Basic Quality Tools," *HRMagazine,* pp. 81–83, Jan. 1992.
4. Burr, John T., "The Tools of Quality. Part VI: Pareto Charts," *Quality Progress,* pp. 59–61, Nov. 1990.

5

DEPLOY:
MAKING ORGANIZATIONAL
IMPROVEMENTS HAPPEN

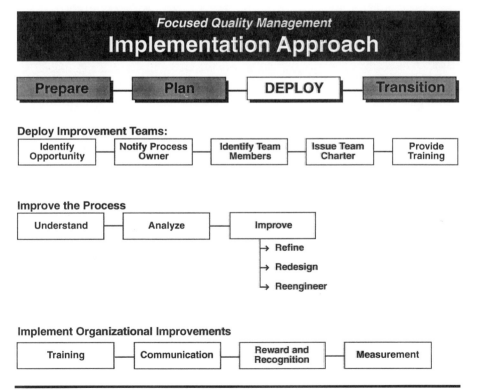

FIGURE 5.1

The second half and the culmination of the *deploy* phase is the integration of organizational improvements to enable processes and people to meet strategic objectives. Organizations improve as their core business processes improve. In reality, however, it is the people in the organization who identify, shape, and implement the changes to those processes. They also are the primary sources of ideas and support for the essential activities that make quality a way of life, whether it be for products and services, corporate culture, or the bottom line.

However, if there is one thing that would guarantee the success of Focused Quality Management (FQM), it is ongoing and effective quality improvement support structures. Without them, people lack the skills, motivation, and feedback required to initiate and sustain meaningful change.

Four important support structures (illustrated in Figure 5.1) go hand-in-hand with process improvement:

> "**W**here corporate culture is strong and robust a distinctive ethos pervades the whole organization: Employees exude the characteristics that define the mission or ethos of the whole, e.g., outstanding commitment to service, perseverance against the odds, a commitment to innovation, or in less fortunate circumstances, lethargy or a sense of helplessness or futility."
>
> Gareth Morgan[1]
> *Images of an Organization*

- ▪ Training
- ▪ Reward and recognition
- ▪ Measurement
- ▪ Communication

These are not the only quality improvement support structures that a company could or should have, but they are the ones whose presence or absence will largely determine the success or failure of a quality initiative.

In improving an organization's support structures, it is essential to consider them as part of an integrated whole, rather than individually. Further, there is no one proper order for developing them, even though in this book training is discussed first and communication last.

Rather, organizational support structures should be addressed during the same assessment and planning process that is used to identify critical success factors and process improvement strategies. In other words, support structures should be considered as important as the nuts and bolts of the business, because they are really the same thing.

This strong emphasis on support structures can be thought of as the scaffolds and cranes used to build a quality organization. Like scaffolds and cranes, the emphasis will be removed in time, as the organization gains the strength to provide its own support. In the short run, a lot of time is focused on training, incentives, measurement, and communication. In the long run, those things are simply there as part of the way the business is routinely run.

TRAINING:
A MIND IS A TERRIBLE THING TO WASTE

Consider Content

In an organization devoted to FQM, people need new skills—ones associated with the behaviors required for organizational success. These new behaviors include:

■ Process analysis

■ Teamwork

■ Attention to others, principally customers

Identifying training needs is part of the organizational assessment undertaken to begin the process of FQM. The first step is an examination of employees' existing skills with an eye to the core business processes and strategic business goals that control the future state of the organization.

If employees already have the skills needed to implement FQM, they can be immediately assigned to process improvement teams. In most cases, however, they will lack some of the skills they need.

This is because, in most cases, employees do not consider their jobs as part of a process. Rather, they have looked only at the day-

---

Providing high-quality service in every interaction with clients is just one part of customer service training at the Royal Bank of Canada. Trainees are also empowered to improve themselves further by designing a personalized outline of their training goals.

A five-course curriculum is tailored to the company's business strategy and client needs. Training goes on in and outside of the classroom. Courses address delivering basic client services, conducting business meetings with clients, ranking client needs, and developing strategies to meet those needs.

Trainees are required to develop an action plan with their supervisors, explaining their career goals based on what they learned in the program and including a list of objectives addressing their strengths and weaknesses.

The customer service training program was designed to support the business unit's strategies, corporate values, and quality services.

to-day tasks they are charged with. Organizational structures and management styles have actually worked to hide the big picture from the worker. As a result, process thinking is foreign; task thinking is easy: "Just tell me what to do and go away—I'll get it done."

Additionally, relatively few employees have been asked or allowed to participate on a team, particularly in the workplace. Even athletic teams tend to emphasize the development of individual skills. Each player has a uniquely defined place on the team, and team success is predicated on collecting the best group of individual players. However, individual performance-based incentives exist in most organizations. Even in professional athletics, monetary rewards (bonus clauses) are largely based on individual achievements rather than on team success.

Finally, many employees have not considered that they have any customer to please other than the boss. Their approach may be to get the job done and make the boss happy, realizing that to fail in that regard results in trouble. Because of both that narrow focus on satisfying the boss and the hierarchical, functional structure of most organizations, most employees have little concept of process and, not surprisingly, little idea of customer.

Thus, employees need to develop and change their behavior, and that requires training. An effective quality training strategy will be linked with strategic business goals. It should identify which people will serve on process improvement teams and then outline how to give them the appropriate skills, including process analysis skills, such as:

- Flow charting
- Pareto analysis
- Cause-and-effect analysis
- Simple statistical measures

Employees also need to learn team skills. Effective team behavior and teamwork are not natural to most people. In fact, many people are uncomfortable working with a team toward a team goal; they would rather go it alone. These people need to be taught:

- True teamwork
- Consensus building
- Integration of individual goals into team goals

These skills should be reinforced by giving people opportunities for practice.

Any organization's quality training efforts will fail if they emphasize *only* tool skills or *only* team skills. In the first case, employees will have the technical ability to analyze and implement process changes, but the process improvement teams will fall apart because they have not learned how to work together and integrate their work. In the second case, team members will feel good about the team and about each other, but nothing will get done because no one has learned how to analyze processes and implement change to improve them.

How to manage the quality training initiative must be carefully considered. The training may be designed and delivered using in-house resources alone, or all or part of the training material may be licensed. Another approach is to hire a training firm to provide the training materials or conduct the training in-house. In any event, training needs must be balanced against the current capabilities of the staff.

An organization that wants to effectively deal with change must have an education and training curriculum that combines formal classroom teaching with on-the-job application. The best approach involves skills development that includes just-in-time training, along with on-the-job coaching for individuals and teams. This way, people have a better chance to actually apply their new knowledge and skills. Coaching plus training provides the best opportunity to transfer abstract learning to real-life situations.

In theory, just-in-time training is a beautiful concept. People are trained in the skills and behaviors they need at the precise moment they need them. In practice, it rarely works this smoothly. The resources required to train are seldom available on a truly as-needed basis. Only with careful scheduling can training resources be used efficiently and not overburden people with superfluous information.

A leading high-tech organization conducted a study to determine the degree of competency

Classroom

Step 1 Theory—by content experts

Step 2 Modeling—show students concepts through the use of customized examples

Step 3 Practice—allow students to work examples through homework, workshops, case studies, and the like

Workplace

Step 4 Application—after classroom work is completed, students begin a project to apply what they have learned in the classroom

Step 5 Experts coach the students to provide continuous corrective feedback to reinforce the material learned in the classroom

Table 5.1
Degree of Competency vs. Training Approach

After these training steps	Degree of competency per 10 students		
	Knowledge: understanding	Skill: demonstrable	Application: transferrable
Theory	8	1	0
Modeling	10	3	0
Practice with corrective feedback	10	7	2
Adaptation (on-the-job project)	10	9	5
Facilitative consulting (coaching)	10	10	8

achieved using a coaching approach versus the traditional training approach.

What the study found is shown in Table 5.1. This coaching approach was used to develop skills within a major financial organization. Facilitators coached five project teams—three process improvement teams and two process development teams. Because the teams had "hands-on" consulting support as they followed the improvement process, they realized rapid results.

For example, one team was asked to improve the orientation and training of new employees. This team created a process flow of orientation and training for customer service personnel. They designed a training curriculum using survey analysis to identify training needs. Measurements of competency were developed and requirements established that allowed the organization to determine readiness for solo assumption of job responsibilities.

A second team was charged with finding a better new product development process. This team created a new product development process that identified prevention points for "hot spots." They created a process owner (project manager) role to coordinate the various aspects of new product delivery without adding head count.

The other improvement teams looked at shareholder communications, the mail process, and the telephone-answering process in the customer service department. The teams also were trained in total quality management skills and provided with on-the-job facilitative consulting.

The shareholder communications team had immediate success. Their goal was to deliver timely and accurate shareholder communications. To improve the current process, the team used process flow analysis, cause-and-effect diagramming, and value analysis. The results included eliminating waste in four subprocesses and developing a regular schedule for standard communications. A savings of $50,000 in postage resulted, along with a savings of 1500 work hours, worth $100,000 a year.

The team focused on the mail process demonstrated similar results. This team's goal was to reduce mail cycle time. A 13% reduction in cycle time resulted. Using process flow analysis, cause-and-effect diagramming, and control charts, this team almost immediately found not only a way to reduce cycle time, but also identified ways to reduce head count. The cost savings was $52,000 annually. No layoffs occurred because employees were redeployed to other tasks.

Hands-on coaching also helped the telephone-answering process team. The customer service department established a goal of reducing abandoned calls from the current level of 4.2%. Control charts and scatter diagrams led to a change in the way the telephone system was programmed and a clear identification of previously unknown patterns affecting answerability. This team's work resulted in a 40% reduction in the number of abandoned calls.

It was clear that these results were due in large measure to on-site coaching by team facilitators, which made the teams' training more effective. Today, the facilitators are transferring their coaching techniques to their international staff.

Once the areas requiring enhanced skills have been recognized and the approach for employee skills development established, the employees can then be prepared to effectively implement change by providing them with the right education and training and giving them the opportunity to put the new knowledge to work.

Effective team training is illustrated by a coal company that immediately followed management training with work group team training at their location and then immediately put the teams to work on specific projects to improve their work processes.

Rogers and Ferketish, in *Creating a High-Involvement Culture through a Value-Driven Change Process,*[2] have identified some of the key areas in which employees often require skills development to prepare them for their new decision-making roles:

- **Business knowledge:** Understanding the organization's situation, including financial information, customer complaints, and competitive position
- **Technical skills:** Knowing how to perform a particular job, such as soldering, editing, operating heavy equipment
- **Interactive skills:** Enabling all interactions to occur so that values are supported, whether the interaction is one-on-one or in groups or whether with peers, subordinates, superiors, customers, or suppliers
- **Continuous improvement skills:** Looking at outputs, job processes, and inputs and systematically analyzing and improving them, including the use of cause-and-effect diagrams, run charts, Pareto charts, and similar tools
- **Leadership skills:** Coaching and reinforcing positive performance, leading effective meetings, rescuing difficult meetings, delegating authority, resolving conflict, getting others to buy-in to ideas
- **Team skills:** Participating in meetings, working on teams, valuing differences, reaching team consensus
- **Service skills:** Encouraging customers to identify the kind of service they need, communicating effectively with customers, dealing promptly with customer complaints

Not all the necessary skills are best learned in a formal classroom setting. Some skills are better learned on the job, through special assignments, or through job rotation.

Planning is the key. Before beginning, and as the process continues, it is important to:

- Understand the change process and the resources it requires
- Grasp the skills necessary for implementing change
- Identify the appropriate forum for skills improvement training
- Carefully monitor the results as people put their training to work

Success should not be measured in terms of numbers of people run through the training program. Improvement efforts have failed simply because they were not effectively planned to take the best advantage of newly acquired skills and behaviors. Nothing is more unfortunate than for a group of employees with newly acquired skills, a high level of enthusiasm, and a vision of what must be done to slowly lose the skills, the spirit, and the vision because they could not put them to work.

Training must also develop—in *all* employees—an appreciation for customers, the people that the process serves. The focus must be directed outward, not inward, so that employees can work on those things most critical to customers. Easy, cosmetic changes that have no effect on customer satisfaction and the bottom line are of no value.

It is easy to assume that customer recognition and customer sensitivity are perhaps more critical in front-line service areas than in metal-bending manufacturing areas, but, in fact, all processes have internal or external customers. To sustain success, it is necessary to concentrate on the critical success factors that satisfy the needs of all customers. Whatever training is provided **must** ensure that outcome.

Dr. Lynn Ward, Director, TQI Education and Support of the Veterans Health Administration, was asked what she saw as issues in training for process improvement. She stated:

"Probably the biggest issue is that TQM seems so simple that some people feel training is not needed. However, training helps establish commonality and consistent focus for the team members.

"A problem in the health care environment is that many people are already trained in components of the process but do not have the big picture. A training program can reinforce that component knowledge and validate the professional role while giving them an understanding of the whole process. Starting without training usually means starting over."[3]

Timing Is Everything

Just as important as the contents of the training is the timing of the training. Successful quality training processes start at the top of the organization and cascade downward. This approach ensures that every layer in the organization is ready to support the change process before teams set out to improve the process.

In a pilot roll-out strategy, the involvement of upper layers of management is focused to ensure that the same support is present before the pilot projects begin. However, the scope is lessened because not every part of the organization is involved. This approach is shown graphically in Figure 5.2.

In any event, quality training is like fine wine: it must not be served before, or after, its time.

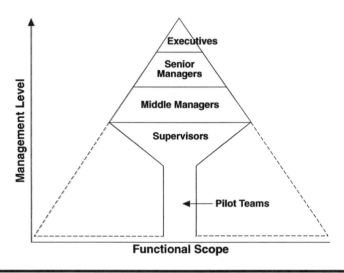

FIGURE 5.2

REWARD AND RECOGNITION: CREDIT WHERE CREDIT IS DUE

People often speak of reward and recognition in the same breath, but actually each accomplishes a different objective. A formal **reward system**—compensation and benefits—is intended to give employees the means for acquiring the fundamental necessities that they and their families require. The premise is that the organization trades a paycheck for their presence and level of effort expended in meeting the basic needs of the organization.

A **recognition system,** on the other hand, is intended to give employees opportunities to be commended, either as individuals or as teams, for extra effort and accomplishments in support of strategic business goals. Thus, recognition systems focus on publicizing success wherever success is found. As opposed to the monetary benefits that most reward systems offer, recognition may take the form of awards or prizes.

Reward Behavior That Supports Strategic Goals

People need more than training in order to function effectively in a corporate environment of FQM. They must be encouraged, in formal ways, to adopt the behaviors and techniques that they have learned in order to achieve organizational goals.

They will do so more readily when the reward system is structured to compensate behaviors critical to organizational success. Like it or not, employees behave exactly as they are compensated to behave.

Success depends on treating people as *people,* not machines, regardless of their position in the corporate hierarchy. A formal reward system—a paycheck—may not be the most effective motivator. Without the formal reward, however, there will soon be no one left to reward. Therefore, a note of thanks cannot be substituted for a check to the bank. Both are necessary.

Rewards, then, must be consistent with the goals of FQM. In other words, employee compensation may need to be restructured in nontraditional ways if the compensation system is to reinforce non-traditional objectives.

Traditionally, wages and salaries are inflexible. Below the executive level, compensation is essentially a fixed cost. Of course, management can reduce hourly compensation costs somewhat by cutting back on overtime, but the only way to find significant relief is to reduce the work force. Such action causes organizational turmoil, carries hidden costs, harms employee commitment and morale, and is incompatible with the principles of quality management.

Organizations seeking to meet the competitive challenge by creating a high-commitment, high-involvement atmosphere in which the concepts of total quality management and process efficiency are practiced cannot ignore the issue of employment security. An employee cannot be expected to feel a sense of partnership with management when he or she suffers a loss of livelihood with every dip in the economy.

Organizations that rely exclusively on fixed pay are severely limited in their ability to weather a business downturn without reducing their work force. They simply have no other relief from significant labor costs.

Perhaps the most serious failing of traditional compensation approaches is that little reinforcement occurs for improving performance. Rewards clearly influence behavior, and if organizations do not explicitly reward the behavior critical to their success, they should not be surprised if that behavior fails to occur.

In today's organizations, individual increases have nothing to do with merit; instead, they reflect such things as cost of living changes, the individual's point in the salary range, budget contracts, and political considerations. In addition, the merit system typically makes little distinction between high and low performance.

The non-reinforcing nature of reward systems is even more pronounced at the level of hourly workers, where pay increases are normally based on seniority, dictated by the union contract, or applied across the board. Hourly employees literally have no incentive to improve productivity, quality, or organizational performance.

Traditional compensation systems also reward conformity or doing things the "right" way. The objectives of FQM are to challenge the status quo and to continually seek better ways to do things.

Of course, this does not mean that anarchy should prevail, with employees doing whatever feels right to them at a particular moment. Rather, teams of people following rational guidelines should plan, structure, and develop the approach to change. Therefore, compensation systems that reward teamwork, rather than individual prowess, must be designed.

Team-based compensation systems that tie individual rewards to team success must become the norm, and not the exception, if real change and real quality are to come about in the organization. Employees should expect to see their salaries tied to group performance in the main, with individual differences based solely on individual contributions to group goals.

Functional skills are important, and functional success is expected. Key process improvement is, more often than not, cross-functional improvement. As a result, compensation systems must encourage cross-functional thinking.

Pay for performance is not a new idea. In the past, however, it has gotten confused with piece rates—systems of paying workers for the amount of output produced. All too often, piece-rate pay was based on quantity to the exclusion of quality, and the ideas of process improvement and customer satisfaction were lost along the way.

One manufacturing organization paid workers by the piece to produce defective and non-defective parts alike. It also paid a premium for overtime to repair the defective parts. The compensation system actually encouraged low-quality work, and the workers collectively followed the dictates of their pocketbooks. Today, that company is struggling to stay alive.

To be rewarded, success must have an impact: team performance must be measured against the overriding goal of customer satisfaction. There must be measurable improvement in customer satisfaction. Where teams effectively identify and implement process improvements that increase customer satisfaction, bottom-line results will follow.

Also necessary is a new management attitude: listening rather than telling, results rooted in quality rather than quantity. Most important is management's ability to instill the belief that bottom-line results will follow from FQM. This requires an unswerving commitment from upper- and mid-level managers. It also requires their willingness to be subject to the same reward system as the rest of the work force.

The compensation system most often mentioned in any discussion of pay for performance is gain sharing. In its simplest terms, gain sharing is a reward system that ties compensation directly to performance that increases the profitability of the enterprise. Employees learn to critically examine their processes and procedures, improve the operational effectiveness of those processes and procedures, and share in the productivity gains for the organization.

According to John Belcher in his book *Gain Sharing:*

> Gain sharing is a compensation system that is designed to provide for variable compensation and to support an employee involvement process by rewarding the members of a group or organization for improvements in organizational performance. Gains, as measured by a predetermined formula, are shared with all eligible employees, typically through the payment of cash bonuses.[4]

Belcher, formerly with the American Productivity & Quality Center, emphasizes that gain sharing offers group incentives that involve current payouts. Because it is based on real gains in bottom-line performance, gain sharing is self-funding.

Gain sharing is consistent with the principles of participative management. Perhaps most important, gain sharing, when properly applied, builds individual accountability toward organizational goals.

Regardless of the compensation scheme chosen, whether it includes pay for knowledge, small group incentives, lump sum bonuses, two-tiered plans, or gain sharing, it must apply the following overriding principle: **reward behavior that supports strategic business goals.** Reward mechanisms should encourage creative thinking, but, at the same time, they must encourage people to channel their creativity along lines that ultimately benefit the customer and the organization.

Many organizations have adopted formal recognition processes for excellence in quality. For example, in 1990, faced with the prospect of international competition and the need to institutionalize the continuous improvement process, VITRO Corporativo of Monterrey, Nuevo Leon, Mexico, sought a world-class reward and recognition system that could be applied to the entire organization. CEO Ernesto Martens assigned the quality award effort to Mario Garza, Corporate Director, Human Resources.

The VITRO Award Team coordinator, Jorge Parada, began with an organization-wide awareness and education campaign that exposed all operating divisions to the application process as well as the award criteria. (The criteria drew upon both the Mexican and U.S. National Quality Award criteria, along with specific unique company requirements.)

In the following months, divisions nominated finalists for the award. The six finalists were visited by a panel comprised of prominent Mexican businessmen and AQPC-CG personnel. In June 1992, the VITRO Quality Award ceremony took place as VITRO's retired chairman, Adrian Sada Trevino (for whom the award was named), bestowed the award on the VITROFLEX/Flat Glass Division. With more than 500 attendees representing all levels of the organization and local media looking on, a historical recognition of excellence in the quality improvement process was made at VITRO. The award continues to be a powerful incentive.

Recognize Behavior That Supports Strategic Goals

Everyone is motivated by the most important of unfulfilled needs, and all people have similar needs. Whereas reward systems aim primarily to fill basic and security needs, recognition systems aim primarily to satisfy belonging and ego needs.

Non-monetary and inexpensive awards, prizes, or tokens are often all that it takes to fulfill the objectives of a recognition effort. When such items are used to recognize special contributions, they are visible and immediate. Everyone in the organization recognizes what the item symbolizes.

In an effort to promote understanding of business objectives at all levels and to encourage people to focus on those objectives, other companies have adopted internal quality awards modeled after the Malcolm Baldrige National Quality Award. Such programs are designed to recognize those organizational departments or divisions that have most consistently fulfilled the provisions of the award.

Blindly copying the provisions of the Baldrige Award may not be effective. Instead, any program should be tailored to the individual needs and culture of the organization. Just giving a prize that resembles the Baldrige Award does not work. Rather, any award needs to be carefully structured so that it truly reinforces the behavior that supports strategic business goals.

For example, the Conference Board found that Xerox Corporation held its first "Teamwork Day" in one location in 1983 and has since expanded the event to 17 locations worldwide. Further, half the respondents in the survey treated recognition as more than reinforcement of quality-related behavior and achievement. Those innovative companies used recognition as a self-assessment tool, involving company teams or other groups in the process of measuring their achievements against a set of pre-established criteria for quality, such as the Malcolm Baldrige National Quality Award or internal criteria established by the individual companies.

> **M**any companies have found that annual recognition events help reinforce employee effort over and above simply doing the job. One company structures an annual event so that about 75% of all employees or teams are recognized for a special achievement of some kind.

Experts suggest that effective rewards must be of value to the recipients. Not surprisingly, the most effective way to find out what employees value is to ask them! This is vital because, in the end, recognition programs are only effective if those who give the recognition understand how to reinforce employee behavior. As always, strong leadership is a prerequisite for success.

> **B**oise-Cascade Timber and Wood Products used their November 1990 Total Quality Leadership Conference to showcase initial process improvement team activities and results. The smiles on the faces of team members receiving a standing ovation clearly demonstrated the power of saying, "Thank you." The impact was obvious when a plywood mill worker asked for the microphone to say, "We the hourly workers would like to thank you for letting us fix our processes: things we knew all along but just couldn't get across to you!"

Non-cash recognition systems are especially important in the early stages of the FQM process and remain important as the process becomes a part of the organization's culture. However, as the FQM process matures, cash incentives and shared savings become increasingly important to non-management employees.

Whatever the stage of the FQM process, studies have shown that compensation tied to quality performance remains the most powerful incentive for upper management. Most importantly, recognition should be memorable. Banquets, ceremonies, and celebrations are additional examples of effective and commonly used forms of ceremonious recognition.

Today, organizations are trying similar innovative reward programs to give their employees more incentives for high performance:

> The military services have always used recognition as a motivational tool, and their approach to quality improvement reflects that heritage. At the U.S. Air Force Quality Symposium in 1994, teams from all over the globe were flown to the conference to tell their stories. The impact was obvious when the Air Force Chief of Staff, Tony McPeak, a four-star general, and Jay Kelley, the head of the Air University, a three-star general, stood next to a young enlisted man and listened as the young man told him how his team improved a process to get greater readiness and lower cost. One observer noted, "That sergeant is so proud he is going to burst or hyperventilate. Look at his grin!" Interesting to note was the fact the generals did not spend time talking to each other. They went to the troops.

MEASUREMENT:
IF YOU CAN'T MEASURE IT, YOU CAN'T MANAGE IT

To determine the overall effectiveness of an organization, it must be possible to measure success. As a result, a family of measures that ties process improvement to strategic business goals is required. Although this may seem to be common sense, organizations rarely develop the link well enough to assess operational effectiveness.

Further, without a suitable family of measures in place, one that all employees understand, it will not be possible to assess the effectiveness of training initiatives. Training should ultimately lead to a better organization. However, un-

"These companies are successful because they get widespread involvement of their people and they have created a culture of measuring. What I like about these organizations today is that you can visit the chief executive or someone down on the plant floor and they're using the same terms."

Alan Marten, Dean
Johnson Graduate School
of Management
Cornell University

less operational and individual performance can be measured before and after the training, it will be impossible to tell if there has been any improvement.

Reward and recognition systems are no different. To be most useful, they must clearly and coherently relate process, team, and individual performance to strategic business goals.

Measurement, then, is the connective tissue that binds all process improvement. The old saying, "If you can't measure it, you can't manage it," holds true. FQM depends for its ultimate success on the ability to measure progress and to set an accurate course.

Organizational measures of the highest order are those that describe the effectiveness of the enterprise. They include measures of profitability, financial performance, market position, customer loyalty, sales, and growth. Called **results measures,** they provide statistics such as market share, return on assets, return on investment, debt-to-equity ratio, and sales per employee.

Process measures, the second tier in the family of measures, are indicators of process capability. They include measures of process variables such as cycle time, quality, error rate, efficiency, capacity, reliability, delay or wait time, and inventory turns. They are the basis for reward strategies such as gain sharing, and therefore must be directly tied to results measures.

Although process and results measures do not share a one-to-one correlation, they do help to explain causal relationships. Without such an understanding, any improvement strategy is the result of mere opinion and guesswork.

Some managers often consider **benchmarking** to be a measurement procedure that compares their own organizations with other similar organizations in order to identify best-in-class or world-class

---

Upon returning from a Deming seminar, a CEO said, "It wasn't very effective; all I remember is he kept saying: 'How do you know?'"

Three months later he praised the quality guru. He said, "Now when any of my staff tells me we should do something, I say, 'How do you know'"

He went on to relate how this simple question, along with management's involvement, enabled his entire organization to focus on measurement. Once that happened, performance increased. The results are fantastic.

As he told this story, an observer could not resist the urge and said, "How do you know?" The CEO responded, "My organization's results!" Deming would be proud.

performance. To be sure, benchmarking is, at heart, a measurement procedure. However, it is a measurement procedure with a difference: it focuses not only on the results of a particular organization, but on those of world-class organizations and on the business practices that they have used to win their results. It helps to identify what it takes to be world class. Done well, it is a road map to excellence. Measures show the organization how far it is from its goals. Benchmarking tells the organization how to achieve its goals.

FQM, with its emphasis on managed change, relies largely on discovery tools like benchmarking. Without a well-defined and well-understood family of measures, however, change strategies cannot be identified or assessed, and, in the end, the changes implemented will prove ineffective.

COMMUNICATION:
THE MESSAGE IS THE MEDIUM

As in any endeavor that involves human beings, the life blood of FQM is communication, within and outside of the organization. For change to occur, vision, goals, and expectations must be clearly communicated.

Numerous organizational surveys taken at the beginning of a FQM initiative demonstrate that poor communication is one of the main causes of employee dissatisfaction. Yet in every organization, it is obvious that communication does go on. The grapevine never withers, and it bears notoriously rotten fruit in the form of rumor, innuendo, inaccuracies, and outright lies. The grapevine fills the void.

Employees want to be effective. They want to contribute in meaningful ways. That makes them hungry for information, and even misinformation will do. Unless company goals are communicated, unless employees are trained in performance improvement methods, unless they understand the family of measurement tools, they cannot direct their efforts appropriately at all times.

The value of leadership and communication is evident when looking at the town hall meetings held by Director Clarence Nixon at the Livermore California VA Medical Center to initiate its total quality improvement process. In a classic display of personal commitment, he, the president of the union, and the director of nursing met with every employee. As Clarence said, "Leadership is not a 9-to-5 job, so if a shift came on at midnight we were there."

The message was simple; Nixon spoke from the heart. He told the employees how he came to be the director, his view of the organization, and what he planned to do. He told them that empowerment could only work if managers and supervisors became coaches and facilitators. He also told them that without their buy-in the process would fail.

He recognized that some people might say such an approach could not be done in a government bureaucracy, but his words hit the nail on the head: "By God, leadership is leadership, there's no damn difference."[5]

Communication strategies should include all available corporate media. Among the most effective are the following:

- Newsletters
- Videotapes
- Town hall meetings
- Informal gatherings
- Quality circles
- Department meetings
- Simply talking with people

The communication that such vehicles relay must consist of facts, not wishes and dreams. As the late W. Edwards Deming so eloquently put it, companies must "eliminate slogans, exhortations, and targets or the work force." What he was trying to explain to management is that people need facts, not meaningless prodding. Therefore, the goal of organizational communication should be to inform, not to manipulate.

Effective communication gets the message across in a factual, not a manipulative, manner. Regardless of the medium, the message must be clear: "We want to improve processes with the help of our employees."

To be most effective, communication should applaud success, both individual and team. It should congratulate individuals or teams who sincerely try to make a difference and use the appropriate tools and methods in their efforts.

The effectiveness of internal communication can be measured and improved if it is treated as a process. First, identify its critical success factors, including quality, accuracy, and timeliness. Then, develop survey techniques to assess how well it has been done, particularly with respect to the most critical success factor: Did people get the message? Finally, adjust communication according to these findings.

The medium cannot overpower the message. Rather than showy tools such as electronic media and virtual reality, organizational communication should center on warm-blooded human beings and *actual* reality (a redundancy, to be sure).

All of the points made here about internal communication apply to external communication as well. When communicating with customers, suppliers, and the public at large, any desire to shade the truth in a more favorable light should be resisted. Instead, recall the adage "truth will out." Organizations that deal fairly and truthfully with the public build relationships that keep customers for life.

> In the words of Howard Cosell, "Tell it like it is." Employees deserve to be accurately informed, with good news and bad. Whatever medium is used, it should always tell the truth.

IN CONCLUSION

Deployment of the organizational improvement plan focuses on the key enablers, those organizational issues that must be addressed in support of FQM. The issues addressed in this chapter are not merely nice things to do. They are the things that must be done if process improvement is to be sustained and quality is to become a way of life in the organization.

Most of the principles outlined here concern an organization's relationships with its employees—how to train, reward, recognize, and communicate with them and how to measure process effectiveness. If employees are not capable, motivated, and aware of process performance, then success is impossible. If they are, they are ready to share in the success of the business.

In closing, let's look at a couple of common myths about people and organizations:

Myth: *People are a company's most precious resource.*

Wrong. A resource is something dug out of the ground, shoveled into a furnace, burned for its energy content, and then swept away.

Myth: *Employees are a company's most valuable asset.*

Wrong again. Just ask any accounting department. An asset is something bought, plugged in, manipulated, and allowed to depreciate over time.

Make no mistake. Employees are people, with ambitions, dreams, expectations, energies, and good will. They respond according to how they are treated, which, in turns, depends on how they are perceived.

Employees who are treated as resources or assets will respond accordingly: blindly, automatically, resentfully, and not too helpfully. Treated as people, they will help a business to succeed: gladly, creatively, enthusiastically, and cooperatively. Whether or not a company acts upon these organizational quality musts, they are a requirement for success.

In conclusion, the objective of organizational improvements is to shape the organization for the future. The objective of both the focused process improvements (discussed in Chapter 4) and the focused organizational improvements, which together comprise the strategic quality plan, is to achieve the company mission and vision. Focused improvements communicate priorities and direction to all employees and provide an organizational road map for Focused Quality Management.

BIBLIOGRAPHY

1. Morgan, Gareth, *Images of an Organization,* Newbury Park, Calif.: Sage, 1986.
2. Rogers, Robert W. and Jean B. Ferketish, "Creating a High-Involvement Culture through a Value-Driven Change Process," *DDI Monograph XVIII,* pp. 1–28 (date unknown).
3. Ward, Lynn, personal communications, Spring, 1994.
4. Belcher, John G. Jr., *Gain Sharing,* Houston: Gulf, 1991.
5. Nixon, Clarence, personal communications, Jan. 4, 1994.

6

TRANSITION:
MAKING IT STICK

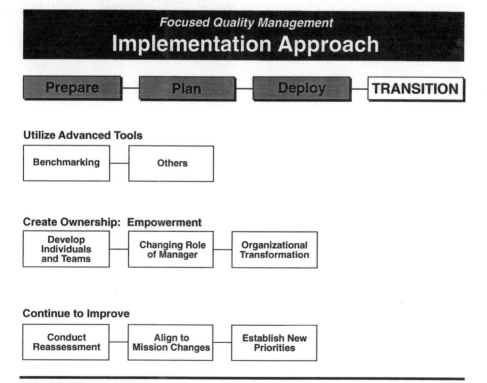

Focused Quality Management
Implementation Approach

| Prepare | Plan | Deploy | TRANSITION |

Utilize Advanced Tools

| Benchmarking | Others |

Create Ownership: Empowerment

| Develop Individuals and Teams | Changing Role of Manager | Organizational Transformation |

Continue to Improve

| Conduct Reassessment | Align to Mission Changes | Establish New Priorities |

FIGURE 6.1

INTRODUCTION

The final step in the Focused Quality Management (FQM) process is *transition* (see Figure 6.1). Transition makes process improvement a way of life for the organization. It demands carefully managed change to continue the shift in culture from defect detection and correction to defect prevention and continuous improvement. Successful transition requires that all employees—not only process improvement teams— have the necessary skills and the authority to do what is necessary to fix processes and make lasting improvements. If the organization does this well, it will truly motivate people to embrace the precepts of quality. If all of the ingredients discussed earlier are in place, the organization will see the positive results that come from the changes that it has set in motion.

Empowered workers are crucial to the *transition*. This means empowering employees to focus on quality every day as a way of life, not just during the quality honeymoon. It requires management behavior to shift from directing employee activities to finding ways to encourage employee behavior that helps to achieve the organization's vision. This requires managers to hone their coaching skills and workers to adapt to their new roles and acquire and/or improve their decision-making skills.

For example, employees who are working as part of an improvement team may have never been in a decision-making role before, and many of them may not have the skills. Telling people they have new responsibilities is easy; however, truly empowering employees requires providing training and creating a learning environment whereby they can acquire the skills needed to make valid decisions. The goal is to have a work force with increased responsibility that knows what is important, does what is important, and manages its own activities while doing it.

During the *transition,* the organization must consider what else it can do to ensure that new processes perform to serve the customers. Improving processes in the *deploy* phase was a step in the right direction, but in *transition,* processes must be monitored, so that they continue to perform as well as possible.

To identify ways to improve processes, the organization may need to utilize some advanced tools, such as benchmarking, to compare the performance of its process to that of processes in other organizations and to learn how it can adapt improved processes to capture the benefits of best-in-class performance. Some advanced tools used extensively in the manufacturing and aerospace sectors are quality function deployment, design of experiments, and other very rigorous ways of focusing on what will satisfy customers and how processes have an impact on that. The purpose of using these is simply to look for opportunities to make things better and keep the momentum of focused improvement going.

Finally, in *transition* the organization must reassess its vision and mission so that they reflect current market conditions. FQM is an iterative process. It is about fine tuning the changes made and keeping an eye to the future by cycling back through the *prepare, plan,* and *deploy* steps to ensure that the organization continues to achieve its strategic objectives and bottom-line results.

Let's take a more detailed look at each part of the *transition.*

APPLYING ADVANCED TOOLS AND TECHNIQUES

"It's the Process, Stupid!"

Benchmarking, as pointed out earlier, is a proven process for identifying, understanding, and adapting outstanding practices from organizations anywhere in the world to help an organization improve its own performance. Specifically, it is the process of understanding what is important to an organization's success, understanding the organization's own processes, finding and learning from other companies that do these processes better than it does, and then adapting that learning to improve the organization's own performance.

Benchmarking is looking for outstanding practices from other organizations in order to learn. The bottom line for benchmarking, just as FQM, is to use that information to improve.

One way to identify outstanding practices is to look at results in other organizations. This can be especially appropriate during the *prepare* phase when conducting an organizational assessment. Benchmarking at that juncture focuses on looking outside, not so much at the processes of other organizations but at their results. This tells the Quality Leadership Team (QLT) how big the gap is between where its organization is and others are. Thus, using benchmarking for gap identification in the *prepare* phase can be a key part of the organizational assessment effort. It is a high-level benchmarking approach focused on performance measures to help executives get a better

"**M**aryland Bank of North America (MBNA) knew that customer service was a critical success factor for their business. MBNA benchmarked customer service at L.L. Bean, IBM, Xerox, and U.S. Sprint. As a result of implementing what they learned, MBNA's industry ranking has risen 38th to 3rd in total annual business done, and profitability has risen 16-fold. Customer defections have decreased to approximately 5 percent. MBNA is now able to provide the fastest line-of-credit increase approval in the industry, averaging only one hour instead of the industry standard seven days."

Carla O'Dell[1]
Senior V.P., International
Benchmarking Clearinghouse

perspective on the existing capabilities of their strategic processes by letting them see how their results compare to those of other organizations.

However, results alone will not yield enough information to change practices. One of the pitfalls of benchmarking is its use as only a measurement activity instead of a process analysis activity. In other words, it is often thought of only in terms of numbers, i.e., comparisons in terms of time, cost, quality, etc. Although such numerical comparisons are useful, even more important in benchmarking is understanding the characteristics of the processes that generate those numbers. Not only is it necessary to know how large the gap is, but also why the gap exists, along with what specifically allows the process to yield outstanding results.

"It's the economy, stupid" was one of the catch phrases of the 1992 Clinton presidential campaign. The Democratic strategists believed that Americans were deeply concerned about the economy and used that as a central theme on which to focus their campaign messages. In the world of FQM and benchmarking, the catch phrase could be, "It's the process, stupid."

Process benchmarking is often conducted by process improvement

"**U**nderstand your own measures and your data, and precisely what you want to learn when you benchmark others. Otherwise, you will be engaging in what has become known as 'industrial tourism.'"

Jack Grayson[2]
CEO, APQC

teams as they search for improvement ideas. At this point, a team has a good understanding of its own process and is ready to gather information about outstanding or world-class processes. The objective is to understand how outstanding/world-class processes operate so the team can use this information to adapt its own process.

It is important to conduct this type of detailed benchmarking only after one's own process is truly understood; otherwise, a lot of money can be wasted on unstructured site visits without a real understanding of what needs to be learned (a.k.a. industrial tourism). One's own processes must be understood in the greatest detail in order to obtain useful answers to such questions as "How do you do what you do?" "What is your flow of work?" "What are your problems and strengths?"

Strategic Benchmarking

Another powerful approach to benchmarking is called **strategic benchmarking,** in which best-in-class and world-class examples of processes that have the most applicability to one's own strategic processes are sought and monitored. This type of benchmarking is usually employed during the *transition* phase. By this time, the process has been improved and is being watched to ensure that it continues to be the best that it can be. Clearly, process improvement followed by employee complacency guarantees that improvement initiatives will not have lasting results. To prevent this, the cross-functional work team responsible for managing the process benchmarks developments in the outside world and stays tuned to those things which could influence how their process could be further improved.

Phase	Who uses benchmarking information	Focus	How it is used
Prepare	QLT	Gap identification	Look at results of other organizations to identify strategic processes for improvement
Deploy	Process improvement team	Process definition	Understand other processes in order to adapt findings to one's own process
Transition	Cross-functional work team · Self-directed · Semi-autonomous	Continuous improvement	Ensure that strategic processes are constantly monitored for ways to improve them

There is always resistance, especially among capable, successful human beings, to the idea that somebody else might be better at doing something. When people are frightened that benchmarking will result in punishment or fearful that they are in fact doing something incorrectly or poorly, they will resist it. Management must take the lead and demonstrate its own willingness to admit that somebody else is better at something, without pointing fingers or assessing blame.

As a matter of fact, ignorance and arrogance are cited as two of the major barriers to successful benchmarking—ignorance of how good other companies are and arrogance about how well things are already done in one's own business.

> **B**enchmarking is the practice of being humble enough to admit that someone else is better at something, and wise enough to try to learn how to match and even surpass them at it.

"The more you benchmark," said James Sierk, when an executive with Xerox Corporation, "the more you accept that other people are very good at things, even better than you. No one company is good at everything. If you look at the best practices from all over the world, you're going to be an extremely good company."[1]

Ultimately, benchmarking, like process improvement, will be judged successful only if it produces results. Positive results do not come from benchmarking alone; they come from successful understanding and implementation of benchmarking findings. What does it take to be successful? There are three primary factors for successful benchmarking.

First, benchmarking must be part of a process improvement initiative solidly grounded in the precepts of FQM. International Benchmarking Clearinghouse surveys show that 90% of the companies that benchmark have an active TQM process. An effective FQM process and infrastructure can help ensure that benchmarking is done in strategic areas and processes and that the findings and learnings are implemented.

> **A**s a result of benchmarking and total quality, Seitz Corporation, a Torrington, Connecticut, manufacturer of plastic rotary and linear motion drive products, has reduced lead times from 16 weeks to four weeks, drawing an average of 220 new projects annually. Process teams of employees were involved both in benchmarking and in implementing improvements.

Benchmarking is a tool for improvement, but it is not the improvement process itself. Without being a part of the FQM approach of *prepare, plan, deploy,* and *transition,* it will not produce results. Clearly, a symbiotic relationship exists between benchmarking and a quality implementation process. Neither one is effective without the other.

The most critical factor in successful benchmarking is the support of senior management. Managers have many demands that compete for their attention. Therefore, just as in process improvement, benchmarking must focus on those processes that are critical to the business in order to sustain the attention and support from management that are necessary to carry the learning from benchmarking to implementation to, ultimately, improvement.

> Robert Toni, President of Coopervision Cilco, a maker of replacement lenses for human eyes, recognized the power of benchmarking. He set out to learn from the best, visiting a dozen companies, including Advanced Cardiovascular Systems, an Eli Lilly subsidiary in San Diego, California. At Advanced Cardiovascular, he found thoughtful, quality-oriented operations management that had created world leadership in medical devices such as catheters.
>
> "When you see it," Toni says, "you say, 'I can do that,' because you talk to the people and they're very proud of what they've accomplished."
>
> Back at his own plant, Toni changed his whole approach to management. "The impact it had in just the first six months on our defect rates, cycle times, and costs was incredible."
>
> In 1986, Coopervision Cilco was number two in the business of replacement human lenses, with a 20% return on assets and an 18% market share. Eighteen months later, Coopervision Cilco was number one, with a 26% return on assets and 22% market share.

Benchmarking and Training

An important factor that must also be addressed to ensure benchmarking success is **training.** Training is just as critical for benchmarking as it is for FQM. Poorly trained benchmarking teams will fail. To ensure that teams are successful and get results, organizations need to institute effective and efficient benchmarking training for teams and provide experts to assist them.

Accelerating the pace of change and searching for breakthroughs are two reasons for using benchmarking. Most organizations do not have the time, money, or energy to reinvent process improvements in a vacuum. Without benchmarking, there is no guarantee that a reinvented process is going to get the kind of results yielded by successful techniques and approaches that others may have already discovered. Benchmarking is also a powerful tool in helping create a sense of urgency to overcome complacency and arrogance.

Most importantly, benchmarking finds those outstanding and innovative approaches, which often come from outside an organization's own industry, that can make it truly successful.

A regional airline was started a few years ago with only three airplanes and very little capital, but a lot of courage. The people who started it knew that in order to make money in the airline industry, they had to keep their expensive equipment working constantly. Airplanes sitting on the ground waste capital and generate no income; airplanes in the air are the greatest sources of both cost and revenue.

One of owners' critical success factors was to fly as many passenger miles as possible. To do this, the airline had to get a plane safely landed, quickly deplane the passengers, clean the plane, refuel, get a maintenance check, board the new passengers, and get airborne. The industry standard for this turnaround time was an hour and 15 minutes. The regional airline knew that if its turnaround took an hour and 15 minutes, it was going to be out of business before it ever got started.

The owners sought advice within the industry, but they could not find any airline that did it in less than 45 minutes. The owners decided that they had to go outside the industry to find the best example of quick turnaround.

They began to look for a similar process, a process in which a piece of equipment under a lot of stress had to be quickly but safely serviced and frequently refueled, its operator attended to or replaced, and the equipment returned to service promptly. The owners looked at a variety of processes before realizing that the process most similar to their own was that of race car pit crews. They decided to benchmark the Indianapolis 500 pit crews, which have an average turnaround time of seconds for fueling, a four-tire change, windshield cleaning, and chassis adjustment.

Because of what they learned from benchmarking the Indy

500 pit crews, their airline now performs an airport turnaround in about 12 minutes by doing all the steps in parallel. For example, the cleaning crew enters the rear of the plane as passengers are exiting the front. The pilots perform their inspection while the fuel is being loaded. Every step in turning the plane around happens concurrently, with airline employees descending upon the plane like pit crews descend on a race car when it comes in to the pit at Indy.

This airline is now one of the most profitable in the United States and has set the standard for others.

It is not uncommon to look outside an industry for breakthrough ideas for making a process the best it can be. For example, Wal-Mart revolutionized retailing with its distribution centers, something it learned from companies like Federal Express, which perfected the hub-and-spokes system. An insurance business is not going to learn effective customer service (i.e., the way to answer a telephone and handle customer inquiries) by looking at other insurance companies. It is going to learn that from an innovation such as General Electric's answer center. Even if a company works well, it will not delight its customers by just doing it as well as everyone else in the industry does; customers consider that level of performance a given.

Benchmarking, like improvement, must be continuous. By the time an organization reaches its current outstanding level, more than likely the level has moved higher. Even if an organization is temporarily the absolute best, limiting factors are always at work, and others are sure to be trying to gain. As Somerset Maugham said, "Only mediocre people are always at their best." Benchmarking must never stop.

Quality Function Deployment

Quality function deployment (QFD) is another tool that organizations can use during *transition* to find additional ways to improve their processes. Technically speaking, QFD is a product/service development technique that pays special attention to customer wants and links those wants to process requirements. Like benchmarking, QFD is a logical, focused, and disciplined thought process that can be used to develop and evaluate the performance criteria for a new product or service and to identify activities that do not contribute or are considered wasteful. QFD is a highly structured method in which customer require-

"QFD is a tool which is used to help translate the voice of the customer into product design characteristics and ultimately to manufacturing to insure that the final product (which may be hardware, service, or any other type of product) meets or exceeds the original customer expectations. The 'customer' here could be the buyer of the product, user of a service, an OEM which purchases parts from a supplier, or any other entity which is a customer for a product."

Ruth Bardenstein and Gregory Gibson[3]

ments are translated into appropriate technical requirements for each stage of product/service development and production.

As illustrated throughout this book, strategic objectives can be achieved by improving existing processes, by refining or redesigning an existing process, or by reengineering a new one. Although QFD is a technique that is especially useful when creating a process for a new product or service, it can be effectively used to help focus on the most important parts of any process as improvement develops. The benefits of QFD are that it assures that products or services are designed so as to maximize the satisfaction of customers, provides a competitive edge, and checks that all critical issues have been addressed in the planning, process development, and improvement activities.

An in-depth discussion of QFD and other powerful tools such as Design of Experiments (DOE) is beyond the scope of this book. However, depending upon a company's specific challenges, these tools can be useful means to help meet organizational objectives.

EMPLOYEE EMPOWERMENT

By the time an organization reaches the *transition* phase, a lot of things have changed and a number of people have been working hard to implement FQM. In the very beginning, a QLT developed mission, vision, and values statements. During the assessment, input from employees concerning quality was gathered. External customers were surveyed concerning quality issues as they saw them, and the performance of major processes was rated. This provided the input to develop a strategic quality plan that addressed both process and organizational improvements.

The next step was implementation. Leadership chartered and launched two different kinds of employee teams: process improvement teams, to work on major cross-functional processes within the organization, and organizational improvement teams, to address organizational enablers necessary to support a quality environment. All of these activities are significant milestones in the quality journey. Step by step, they help to build a framework that facilitates the transition to a high-performance organization.

All of these steps require **empowerment.** While the QLT was empowered to guide and facilitate the quality journey, assessment teams were given the responsibility and authority to evaluate the current performance of the organization in several key areas. Improvement teams were empowered to take the actions needed to understand, analyze, and improve processes.

"Empowerment of us [Federal Express] means our employees know they have the power to make the decision to do whatever it takes to meet or exceed the customer's requirements."

Roy Golightly[4]

In this *transition* step, the issue is not whether or not to empower employees. Empowerment has already happened. Rather, the issue is what *form* empowerment will take as it spreads throughout the organization and becomes a way of life. What, specifically, should empowerment look like in the organization?

"In this recent book, *The Ultimate Advantage: Creating the High Involvement Organization,* Lawler argues that in some situations empowering employees may be doomed to fail. Where pay is low and turnover is high, for example, making the necessary investment of time, money, and training to provide employees with the decision-making tools they'll need may not pay off. That's often the case for service industries, such as fast-food restaurants, banks, convenience stores, large discount stores, and telemarketing businesses."

Kathleen Cahill[5]

The term *empowerment* evolved out of the employee involvement movement of the late 1970s. Empowerment means giving individuals and teams responsibility for the processes with which they are involved. It offers the organization increased responsiveness to customer needs by giving employees more authority to make process changes as needed to fulfill those needs.

It is just as important to know what empowerment is *not.* Empowerment is *not* unplanned. Any rush to empower employees without careful thinking and good planning results in chaos. Empowerment is *not* industrial anarchy. Rather, it is a team effort with managers and other employees working together. This means no one has carte blanche to do as he or she pleases. There are clear limits and well-defined accountability. Empowerment is supported by effective coaching and mentoring to give people the tools needed to exercise their authority and carry out increased responsibility. As a result, empowerment is *not* static. As individual knowledge and skills increase, areas of responsibility expand to capitalize on that greater capability.

There are many definitions of empowerment, but all convey the same basic idea:

> Empowerment is delegating responsibility, authority, and accountability to front-line levels in the organization, where responsive action needs to be taken to satisfy customer expectations.

When this concept is implemented in an organization, performance capacity of individuals and work teams is maximized. Inherent in this definition are certain requirements, including:

1. A coaching process that prepares individuals and teams to perform at higher levels of responsibility and authority.

2. Formal support systems, such as policies and procedures that reinforce responsibility and authority at appropriate levels, reward systems that encourage responsiveness and problem solving at front-line levels, etc.

3. Ownership by the individuals and teams accountable for the results of their actions.

Successfully developing an empowered work force requires a sound strategy. In carrying out their strategies, some organizations have used participatory management and employee involvement as their primary catalysts for change in core processes. Others have taken an approach that reflects the belief that an empowered workplace is an important strategy in its own right. The largest number of organizations have moved toward empowerment and greater self-direction as a result of their quality improvement efforts.

Experience shows that the quality process is an excellent springboard to an empowered work force. FQM gives an organization the opportunity to make a successful transition. Whatever the motivation for and method of transformation, empowerment has proven to be an excellent tool for managers who hope to improve organizational performance.

In a recent *Harvard Business Review* article, restaurant entrepreneur and author Timothy Firnstahl describes how he empowered his front-line waiters and waitresses, but failed to empower his managers. The results of the managerial empowerment he implemented to correct his oversight are impressive because as empowerment shifts responsibility to lower levels, middle managers may be left with little authority. Senior management must pass to them the same kind of increased responsibility if empowerment is to be successful. As Firnstahl explained:

> Several years earlier [late 1980s] I had empowered our front-line employees with a program and company credo called WAGS—*We Always Guarantee Satisfaction*. It gave food servers the authority and power to make immediate amends to customers for slow service, confused orders, overcooked meat, that sort of thing, by giving away food or drinks or even picking up a check.
>
> Now it suddenly occurred to me that I had failed to extend this same kind of power and responsibility to my local managers. WAGS had done wonders to increase repeat business and employee commitment and, by proactively identifying problems before customers even had a chance to complain about them, to help us find and correct the causes of each system failure.
>
> But where food servers had the power to take action on their own initiative, my line managers—bar managers—were still firmly subordinate to corporate headquarters when it came to payroll, menus, marketing, product development, accounting, training, hiring, and general decision making, just about everything except reservations and breathing. We were still operating under an antiquated line-staff system as obsolete as the centralized communist approach.
>
> To replace the biggest staff functions of all—planning, problem-solving, and decision making—I began picturing a weekly meeting where managers would discuss company headaches and opportunities. I wanted them to see problems not as mistakes, a word that puts people on the defensive, but as targets for everyone's input. Our fundamental rule would be, "No one is wrong or right, everyone is a resource. Including me." The resource concept would be the foundation of our weekly meet-

ings. I decided to call this new collaborative process "ally management."

It was ally management that rescued the fresh roasting (restaurant) concept, and I think it can save us from a national epidemic of restaurant deaths. It was also ally management that solved the cost problem I'd struggled with in vain for 20 years. I'd tried bonuses, praise, threats, outside consultants, classes, tapes, videos, and seminars. Nothing had worked very well for very long. Now without Herculean effort, we've achieved the best control results we've ever had. From 1991 to 1992 cash flow went from negative to positive for 200% turnaround, and I've recouped more than half the cash lost during our negative cash flow period.

I see quicker company change, more powerful motivation, and a much more confident and hopeful culture. We have achieved the best cost control numbers in the history of the company. We have reduced the time it takes to produce end-of-the-month financial statements from ten days to three. We have done away with our bureaucracy and turned business around in a down market. In the process, we have reestablished a basic capitalist principle: making money. All because our managers are managing themselves.[6]

There are two major types of empowerment. Both are required if an organization is to share responsibility and authority at all levels. The first is self-empowerment, by which employees, believing in their own abilities and wanting to grow, accept increased responsibility for their actions. The second is a management philosophy and strategy that actively works to improve performance by empowering all employees to improve. The first, self-empowerment, is not something management can impact directly. However, there are many personal development seminars available that management can offer to employees who want to enable themselves to do more.

Creating an empowered workplace enables all employees to work together to make the transition to FQM as the organization's approach to day-to-day business. This requires:

1. **Understanding the development needs of both individuals and teams to prepare them to effectively self-direct their activities.** FQM can only be achieved when organizations have

developed a solid understanding of these needs. During the *deployment* step, organizations train employees, especially front-line employees, to give them the tools to improve process capability for meeting customer needs. Process improvement teams with the responsibility and authority to improve key processes have begun to get results. *Transition* builds on these activities. It takes what has been learned by individuals and teams in the first implementation initiative and starts to focus on doing what is needed to expand it to the organization as a whole.

2. **Understanding the changing roles of managers as the organization moves toward self-direction.** Many organizations will encounter problems developing coaching and mentoring skills essential to the changing roles of managers in FQM. Helping managers to accept these changing roles and be successful in them can be a challenge. Again, during the *transition* the organization builds on what it has already learned and pulls it together into an integrated whole.

3. **Designing and structuring the organization's transformation toward self-direction.** During the *planning* step, the organization laid the foundation for organizational change. During *deployment,* it chartered and launched organizational improvement teams to address communication, training, reward and recognition, and measurement issues. Again, the *transition* step builds on these efforts and continues the redesign of the organization so that FQM indeed becomes a way of life.

An impact on bottom-line results in the FQM-based organization requires committed individuals, often working together as teams, who are trained and motivated and willingly contribute their expertise to produce the work best suited to achieving the organization's goals. Communication, training, rewards and recognition, and measurement systems are firmly established during the *transition* phase to make this a reality. Empowered employees know what is needed, are trained to accomplish the task, are rewarded for doing so, and are measured on real outcomes.

As stated previously, during *transition,* the organization puts in place the processes that enable it to achieve its vision. The scope of the self-direction that can be allowed is usually a consideration at this point. It is important to note, however, that the goal is not a self-directed work force. Rather, the goal is a work force evolving toward increased responsibility and authority at all levels in order to increase the performance capability of the organization. Self-direction *can* be an outcome, and increased empowerment and shared responsibility are the milestones along the way. The ultimate goal, however, is achievement of strategic goals.

The QLT must be prepared to make a sound decision about the degree of empowerment that is right for their organization. Experience shows that a road map of the various stages of empowerment helps. A continuum of employee involvement, illustrated in the following section, shows how organizations move from a traditional to a self-managed culture.

Most organizations begin involving employees with simple initiatives, such as suggestion programs or employee attitude surveys. As an organization matures, parallel structures, such as task forces, leadership teams, process improvement teams, organizational improvement teams, and other types of teams, are created, which elevates employee involvement to another level. While this has been the thrust of the *deploy* step of the FQM process, it represents only an interim step. To support full empowerment in a successfully self-directed workplace, the entire organization and all its processes must be integrated into a management-led, worker-supported, customer-focused endeavor. The job of management is to make this happen.

TOWARD PROCESS EMPOWERMENT

The movement toward **process empowerment** is characterized by modeling, analyzing, and improving business processes. Most organizations either evolve into business process empowerment by implementing increasingly sophisticated versions of problem-solving, or they leap into it by reengineering or significantly simplifying a business process. Both of these approaches utilize a combination of empowerment and Total Quality (TQ) techniques, as follows:

- **Align:** Establish alignment using quality function deployment, policy deployment (Hoshin planning), or automation deployment
- **Frame:** Describe a process for process identification and boundary testing
- **Model:** Select key processes and build a model

- ▪ **Analyze:** Quantify the gaps between customer requirements and current process outputs
- ▪ **Choose:** Select a few methods of improvement out of the library of TQ improvement techniques
- ▪ **Remedy:** Brainstorm, select, and test remedies to various root causes
- ▪ **Measure:** Develop solutions, implement them, and track progress using specific measures built into job assignments and evaluations
- ▪ **Replicate:** Copy aspects of process improvements into other operations

According to Richard Greene in his book *Global Quality,* an increasingly popular tool for process empowerment is a process empowerment room, where computer information appliances perform knowledge and data delivery in ways that achieve behavioral change. Process empowerment rooms are a completely integrated, future-state architecture for team activities throughout organizations. They combine personal digital assistants, portables, workstations, electronically equipped conference rooms, and overall computer network coordination. They are administered by two facilitators: one facilitating the group process and the other the integrated software environment. These rooms allow the process team to view the future of knowledge-enhanced products and processes. The result is an increase in futuristic thinking, visioning, and positive morale.

Once an organization has learned how to successfully implement a variety of quality improvement team initiatives and has experienced its power at a variety of levels within the organization, a question still remains: How much further along the continuum should

"**A**t General Electric a program called 'Work-Out 'evolved from a mandate from CEO Jack Welch: 'We've got to force leaders who aren't walking their talk to face up to their people.' The result was an ambitious, 10-year effort to spread a new way of thinking and acting to all levels of the corporation.

One of Work-Out's major goals was employee empowerment. 'The people closest to any given task usually know more about it than their so-called superiors. To tap workers' knowledge and emotional energy, the CEO [Welch] wanted to grant them much more power. In return, he expected them to take on more responsibility. 'There's both permission and obligation,' he says"

Noel M. Tichy and Stratford Sherman[7]

the organization travel? Some decide natural work units are the most suitable framework for empowered employee activities in the organization. Others seek to set up semi-autonomous work teams. The choices are many. To make these decisions effectively, management needs to know the consequences of moving further along in the process. Specific actions, such as developing individuals and teams, facilitating the managers' changing role, and appropriately structuring the organization, are all essential.

Enabling employees to assume increased ownership responsibility and accountability should be one part of an effective quality improvement initiative from the very beginning. However, continually developing and enabling individuals becomes a mainstay management task, because most employee involvement efforts have been focused on chartering, launching, and enabling teams. Developing work teams is a newer focus.

Few managers would argue that they do not have a responsibility to develop individuals or teams. Unfortunately, few use direct teaching or coaching, sharing information and experience, discussing lessons learned, and other supportive mentoring techniques on an on-going basis. A large gap exists between what managers know they should do and what their employees see them do. The answer to the question "Does your boss coach you in ways that help you to perform better?" is usually "No!"

Managers must improve their coaching skills; it is critical for peak performance. A learning environment is vital to successful coaching. It is also essential that work groups share a sense of mission and purpose. Learning how to give supportive, behavioral feedback is key. Being concerned with not only what has to be done, but also how it is done (the process), opens up a myriad of opportunities for growth. Learning to look at problems as opportunities also helps. Finally, creating an atmosphere in which effective management–labor relationships are expected and worked toward as a basis for excellence is essential. Coaching and

"**F**rom day one, Federal Express has had to give its front-line employees wide decision-making power to keep its guarantee to customers: *Your package is delivered on time, or you don't pay.*

Then, as now, Federal Express found that communication was absolutely imperative. And today, it has invested billions of dollars in high-tech communication systems and communications techniques."

Roy Golightly and Jean Ward-Jones[4]

empowering individuals is a two-way street. In return for their coaching efforts, managers receive employee commitment and high performance.

When managers are called upon to share responsibility and authority with employees, they often feel insecure. That is not surprising, because they know that the organization still holds them responsible. "The buck stops here" is still a very real part of their work life. Yet, by deciding to entrust team members with the responsibility for the work unit, managers are making a decision that results in increased performance. That decision is easier to make when teams have proved to be successful during the initial implementation.

Another key to making the decision easier is a strong communication link between the team and the manager, so that the manager can review performance on a regular basis. Creating this communication link should be part of the initiation activities of the team.

Self-Directed Work Teams

For many organizations, **self-directed work teams** are the next step in the quality journey. A self-directed work team is simply a group of employees who have responsibility and accountability for a whole process. The team is called self-directed because it has full authority to direct its own work in a way that delivers products and service that are acceptable to the customer.

This type of team plans the work and performs it, fulfilling many of the tasks traditionally done by supervisors. For example, the team may determine its own schedule, set goals, give performance feedback, hire, and even discipline or fire. As in any development process, the team is not expected to perform all these tasks from the beginning. The team's duties and authority grow as it demonstrates increased maturity and skills. Managers must coach and mentor these teams just like any other team to help them grow.

> "'**M**y view of the 1990s is based on the liberation of the workplace.' He [Welch] said, 'If you want to get the benefit of everything employees have, you've got to free them—make everybody a participant. Everybody has to know everything, so they can make the right decisions by themselves.'"
>
> Noel M. Tichy and Stratford Sherman[7]

Self-directed work teams generally share the following characteristics:

- Members possess a variety of technical skills.
- They are accountable, at a minimum, for quality costs and schedules.
- Members have the interpersonal skills necessary for effective teamwork.
- The team is constantly encouraged to increase its skills and improve its products or service.

Self-directed work teams are similar to process improvement teams and organizational improvement teams in that customer satisfaction and strategic objectives are their focus. They differ in that:

- Self-directed work teams consist of multi-skilled and cross-trained employees with *on-going responsibility* for a complete process. They are held accountable not only for improving their process, but also for continuously producing specific results.
- Quality control and maintenance are part of the team's responsibility.
- The teams schedule which of their members will do what operational work tasks.
- Leadership is shared among team members.

However, in many organizations, the difference between process improvement teams and self-directed work teams is more a matter of degree than substance. An organization's experience with FQM and process improvement teams gives it a solid foundation to build on. As teams mature, many begin taking on some of the tasks listed above in the course of their improvement efforts. This is why process improvement teams have proven to be such a powerful aid to the transition to self-directed work teams.

Self-directed work teams are being implemented in more and more organizations, because they have shown themselves to be more committed and more responsive to the ever-changing marketplace. A wealth of evidence shows that self-directed work teams can constantly improve the system by keeping data, solving problems, identifying changing customer needs, and using their knowledge to stay ahead of the competition.

Obviously, in order for self-directed work teams to be successful, the team must have management support. There must be top-down commitment to sharing knowledge, information, decision-making, and rewards. As discussed earlier, training must be provided for everyone involved.

Motivation and support from senior management are vital to the success of a self-directed work team effort. The teams must be empowered to act with the authority necessary to meet customer needs. In addition, union and management must form a winning partnership to establish new attitudes and definitions about work and the union–management relationship. Rewards must be realigned to reflect teamwork and improvement. Workers must be assured that their improvement activities will not eliminate their jobs.

It is critical to understand and accept that the movement toward self-directed work teams is evolutionary. Transferring too much responsibility before the team is prepared to accept it is as dangerous as withholding delegation of authority when it is ready. Clear plans and expectations need to be developed in the early phases of the effort and must be communicated to all involved in the change.

Experience with team charters provides a good precedent. The better the charter, the better the process improvement team understood and carried out its task. The more clearly defined the charter, the better the results. The same is true here. The better the plan for developing and deploying self-direction, the better the outcome in terms of successful teams.

Realistic Expectations and Time Frames

If an organization wants to accomplish a successful transition to self-directed work teams, its management must develop a realistic time frame. The transition can take two to five years, depending on the state of the organization. Obviously, the more successful the organization has been in quality implementation efforts, the easier it will be to take this next step. Throughout this time of transformation, several major activities require emphasis:

- ▪ **Training:** Identifying new skills needed by team members, and designing and implementing programs to build these skills
- ▪ **Communication:** Designing and implementing a system to deliver a broader range of organizational information to a broader range of employees
- ▪ **Reward and recognition:** Realigning reward and recognition systems to support self-directed work teams

■ **Management support:** Mentoring managers in their new roles so that they, in turn, can mentor their teams in the new way of life by successively delegating more and more responsibility and authority

Organizational improvement teams will have already addressed several of these activities when they implemented FQM in the organization. Their work will provide an excellent foundation for this next phase of the empowerment journey.

Managers transfer more and more decision-making authority to work teams as they mature and grow from being leader-dependent to being independent. For example, managers retain most decision-making authority while team members get comfortable with understanding their work process and each other. Initially, members of the team are busy gaining the skills necessary to progress to the next stage of their development. Later, managers begin the transfer of simpler tasks, such as work scheduling, to the team. Managers should expect some confusion at this point and reinforce team member initiatives whenever possible. Fear of making mistakes can be driven out as the team sees managers coach rather than scold because of mistakes made while learning.

A more equal distribution of decision-making follows as team skill and confidence and the manager's trust in letting go increase. New communication pathways are developed which facilitate the change. Typically, at this point, managing budgets and costs, as well as responsibility for overseeing the whole process, are transferred from the manager to the team. Roles and responsibilities are becoming clear.

As the process continues, some characteristic things emerge such as cross-training of team members, honing of team skills, and the maturing of the work unit into a self-regulating body. By this time, teams are controlling daily operations and assuming significant managerial responsibility, such as hiring. Because of effective communication, teams are familiar with organizational goals, and because of their experience they tend to manage conflict or performance-related issues within the team. Members now look to their manager as a coach, mentor, resource gatherer, barrier buster, and boundary spanner.

In the long run, teams grow more comfortable with responsibility and authority. During this time, informal peer reviews are replaced by a formal system of performance planning, and rewards and recognition are now managed by team members themselves. In many cases, teams demonstrate both the maturity and the knowledge to discipline and even fire team members. Finally, financial reporting is performed by the team, and the results are sent to management.

Some teams are led by an overall team leader. Many share, even rotate, leadership responsibilities among themselves. As teams mature, customers become an integral part of operations, and partnerships are developed that include information exchanges, well-defined roles and responsibilities, and new ways of working together. A process for reviewing the effectiveness of these interfaces is set in place.

As employees grow with self-direction, the organization's systems and infrastructure must change. Information and reward systems should realign to meet the needs of the new environment. In addition, the broader organization must also realign. This can mean new challenges and changes for support functions such as accounting and personnel. These new challenges confront not only the people working in these functions, but also the very way these functions are organized. In the end, form follows function, and the organization changes to meet its needs.

> "'In seven years,' says Welch, 'people who are comfortable as coaches and facilitators will be the norm at GE. And the other people won't get promoted. We can't afford to promote people who don't have the right values.'"
>
> Noel M. Tichy and Stratford Sherman[7]

ALIGNMENT AND REASSESSMENT

The final step in the *transition* phase is actually the step that starts the FQM cycle over again. The organization reassesses its vision and mission to be sure they reflect current market conditions. It asks: "Have we changed our vision?" "Has our mission changed?" "Are we in such an economic situation that some things are more important now than when we first went through this process?" If any of these questions yields a positive response, plans must be adjusted accordingly.

Because of what has been learned since the organization began the *prepare* phase, it may want to make some changes based upon new knowledge. In addition, in today's rapidly changing environment, the organization must make sure people are still marching in the right direction, and if they are not, make the appropriate course correction.

Organizations also must reassess to determine what impact the changes made are having on customer satisfaction. The question is simple: "What do customers think now compared to what they thought before?"

Based upon this feedback about organizational factors, customer satisfaction, and an assessment of market conditions, the organization may need to establish new priorities. For example, it may need to improve a process that has come to the strategic forefront because of technological breakthroughs or create a new product or service to meet customer needs. It may find that the organizational culture is not as supportive of employee empowerment as it should be and organizational improvements are needed.

Changes needed are carefully laid out and reflected in an updated strategic quality plan. Updating the strategic quality plan requires consensus from the leaders of the organization to ensure they support the new priorities as they did the initial endeavor. Communication is also important, because the next round of quality initiatives must be understood by all employees if it is to serve as a road map for the organization's on-going activities.

As mentioned in the beginning of this chapter, FQM is an iterative process, and the *transition* phase is about just that. It is about fine-tuning the changes, benchmarking to establish the status quo and identify how to improve, using advanced tools and techniques to continue the improvement process, empowering people to excel, and building an organization to encourage, support, and reward superior performance. Because standing still is losing ground, the organization must keep an eye to the future by cycling back through the *prepare, plan,* and *deploy* steps to ensure that it achieves its strategic goals.

The FQM process is now in place, and the first phase of the quality journey is complete. As a result of efforts during the *prepare* step, the organization has a better idea of where it is going (the vision), what it is doing (the mission), and how well it has done (the assessment).

During the *planning* step, the organization learned what things are important (the strategic objectives), how to achieve them (critical success factors), and what makes them happen (the key processes).

During the *deploy* step, the organization improved by changing how it trained, communicated with, rewarded and recognized, and measured its people and itself. At the same time, processes were refined, redesigned, or reengineered to better meet customer needs and strategic objectives.

During the *transition* step, the organization applied advanced tools and techniques to continue the improvement process, reemphasized employee participation and involvement to empower people to meet organizational goals, and reassessed its performance and position to reaffirm, if possible, and realign, if necessary, its strategic direction.

The organization did all these things because *people* made it happen! Some people took it upon their shoulders to start the process. Of all the people that made the changes happen, none was more important than the leader. No one has a more difficult task. The focus of the next chapter is what it takes to be that leader.

BIBLIOGRAPHY

1. O'Dell, Carla, "Benchmarking: America Looks to the Customer and Best Practices," *Continuous Journey,* pp. 6–7, Sept.–Oct. 1992.
2. Grayson, Jack, "Benchmarking: Learn or Die," *Continuous Journey,* pp. 8–11, Nov.–Dec. 1992.
3. Bardenstein, Ruth L. and Gregory J. Gibson, "A QFD Approach to Integrated Test Planning," *ASQC Quality Congress Transactions,* pp. 552–558, 1992.
4. Golightly, Roy and Jean Ward-Jones, "Communicating Empowerment at Federal Express," *Commitment Plus,* pp. 1–4, March 1993.
5. Cahill, Kathleen M., "The Many Faces of Empowerment," *Enterprise,* pp. 26–29, Jan. 1993.
6. Firnstahl, Timothy W., "The Center-Cut Solution," *Harvard Business Review,* pp. 62–65, 70–71, May/June 1993.
7. Tichy, Noel M. and Stratford Sherman, "Walking the Talk at GE," *Training & Development,* pp. 26–35, June 1993.

Part III

LEADING THE CHANGE

Thus far, the emphasis has been on the *why* and the *how* of Focused Quality Management. This section describes the *what:* how a leader makes FQM a reality for an organization.

First, some things some leaders have done as they led their organization through the FQM process are outlined. Second, a short review is provided to help determine if an organization is ready to try FQM. Third, some ideas for recovery are supplied should the effort falter.

7

LEADING THE CHARGE

There is nothing more difficult to take in hand, more perilous to conduct, or more uncertain in its success, than to take the lead in the introduction of a new order of things.

Niccolò Machiavelli

Complex change is never easy. It is natural for all of us, individually and collectively, to resist change and to feel uncomfortable while it is coming about. When we feel uncomfortable, we just as naturally put up barriers against the source of the discomfort. When it comes to improving business processes, and organizations, managers tend to encounter problems because of failings in six main areas:

- Vision
- Skills
- Incentives
- Resources
- Action planning
- Effective measures

Unless a leader can overcome employee resistance to change, improvement efforts are only going to yield less than perfect results.

LEADERSHIP COMMITMENT AND VISION

There are few executives or managers who do not consider themselves committed to achieving their organization's objectives. Yet Focused Quality Management (FQM) can fail primarily because of management's lack of leadership and involvement. Organizations that have successfully overcome this barrier understand

■ The value of leadership
■ The importance of vision

The Value of Leadership

Relevant change never occurs without leadership. For example, a couple of years ago, the CEO and executive staff of a small bank in central Kentucky were forecasting the changes coming in the banking industry and how the bank might prepare itself for making those changes. Up to that point, the bank had succeeded; it had done things very well. "But," the bankers were advised, "you either need to change the way you do business, or you're going to fall on hard times, and it'll happen so suddenly that you won't even realize it."

> "We get in trouble around here when we let our people think."

When their operating committee was interviewed, one senior vice president actually said, "We get in trouble around here when we let our people think."

A year later, the bank had laid off 20% of its staff. The CEO and the COO had retired. Under new management, the bank was a prime takeover target. Business and life will never be the same for those people because they did not understand the kind of leadership required to change from the way they had done business to the way they needed to do business.

> "Leadership is nothing more than the art of finding out which way people are going and then getting out front"

Some people maintain that leadership is nothing more than the art of finding out which way people are going and then getting out front. There is probably some truth in that definition. People usually want to go in the right direction, but they need someone to lead the way.

For example, a young executive took a position as quality director with a bank in Minneapolis. When discussing the job during his interview, the executives at the bank had said, "We think some of the things that happen with quality in a factory can happen in a bank." He agreed.

The executive had worked at the bank for about a week when one of the senior vice presidents invited him to lunch. In the executive dining room overlooking the Mississippi River, the vice president told him: "My biggest quality problem is keeping my people motivated. What can I do to motivate them?"

He replied: "It seems to me that your people were motivated when they came to you. They wanted a job, didn't they? What have you done in the meantime to mess that up?"

The vice president did two things that really impressed the young executive. First, he didn't fire him. Second, the vice president took his comment to heart, telling him that he had never thought that way before. The two then worked together over the next few months.

> "Leadership is the ability to cause others to follow into areas of uncertainty."

The vice president and a group of his people went through training in FQM and process analysis. The vice president led his group as it looked at business processes with an eye to improving them. As a result, the vice president learned it was not that his people did not want to help; they did not know *how* to help. They also did not know how the vice president would take their offers to help. Once he ironed out that problem, the vice president truly became a leader.

What is leadership? William Ouchi provides a good definition: "Leadership is the ability to cause others to follow into areas of uncertainty." We are always faced with uncertainty, and if we can cause others to follow us through those areas of uncertainty, then we are, or can become, leaders. Obviously, the trick is knowing the right direction in which to lead.

The people who put together the criteria for the Malcolm Baldrige National Quality Award understand that leadership is integral to the total quality improvement process. They also understand that individual leadership and organization management are two separate things, as illustrated in Figure 7.1

On the left is leadership which is the *driver* of the quality process and is one of the seven major categories used to judge business excellence. In the center are management categories that make up the *system:*

- Information and analysis
- Strategic quality planning

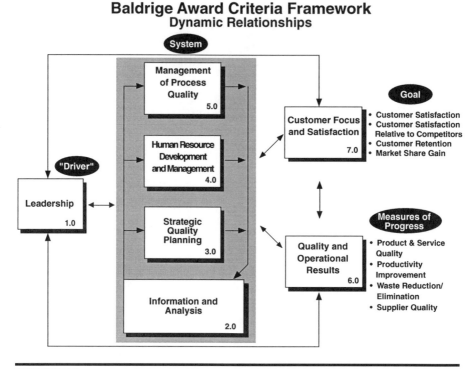

Baldrige Award Criteria Framework
Dynamic Relationships

FIGURE 7.1

■ Human resources development and management

■ Management of process quality

Without both leadership and management, however, business excellence does not happen. People who are good leaders are not necessarily good managers, and good managers do not necessarily make good leaders. That does not make either one a bad person. Both kinds of people are needed for the different jobs that must be done. In some cases, they can even be done by the same person.

About 45 years ago, J.M. Juran started informing the people who run organizations that they have two fundamental tasks:

■ To prevent change when change is not wanted, to manage the organization in a stable and consistent way

■ To create change when change is necessary, to lead the organization to new levels of performance

Those are two different activities. Managers must take care of resources in times of relative stability, and leaders are needed when things need to change. Juran recognized this 45 years ago, and business and industry is finally beginning to discover it today.

> "Management is the art of preventing unwanted change and of using an organization's resources in a consistent and controlled way."

What, then, is management? Management is the art of preventing unwanted change and of using an organization's resources in a consistent and controlled way. Even when managers try to keep things calm and stable, things happen. They need to learn to control those things. Someone once said that management's task was "to have been trained in the organization in the conduct of cattle drives only to be dropped into the middle of a stampede."

Management's goal is to achieve the organization's objectives and to efficiently use its resources. Understanding FQM means that managers need to begin understanding how to look at processes.

In other words, managers must look at more than the organization's results; they must look at

■ Operational processes and procedures

■ The management and administration of resources, human and otherwise

■ The cross-functional business processes that customers experience, even though the organization is not managed that way

If managers understand the business processes well enough, when the organization needs to change, everyone will be working on the right processes. In times of transition, management skills help to direct that change. Leadership skills point the way forward, identify what change the organization needs, and show where it is headed.

There are differences between how managers and leaders operate. For example, John F. Kennedy was able to articulate a "new frontier" for the country, a vision of a "man on the moon in this decade." He was a great leader. Yet Lyndon Johnson managed to push more social legislation through Congress than any president since Roosevelt. Johnson made what he called "the Great Society" happen by working with the Congress. He understood and was a great manager of the legislative process. Like nations, organizations need both leaders and managers.

The Importance of Vision

Effective leaders share several traits:

■ A vision for the enterprise
■ Personal values that exemplify the vision
■ The ability to articulate that vision
■ The ability to motivate people to fulfill the vision

A person missing any one of these traits is not really an effective leader.

What makes leaders? There are two traditional approaches to leadership development:

■ The business school approach
■ The military approach

Business schools spend a great deal of time training managers to accomplish the functions of management. They teach people how to do accounting, how to do marketing, how to do business administration, and a host of other activities, along with some grounding in the principles of organizational development and human resource management. They train them to do specific jobs, hoping that, in times of crisis, leadership capabilities will emerge in these people. Sometimes it works, but often it does not.

A man and his family took a rafting trip down the Colorado River. They ran the rapids through the Grant Canyon for about a week. Their raft had a small motor that did little or nothing against the river's current; it provided little direction. That was the leader's job. As the raft approached a rapid, the pilot's eyes looked far ahead. He moved the motor every once in a while, but his eyes never deviated from the goal. He left the people in the raft to paddle when needed to make sure the raft did not hit the rocks, but his eyes always looked well beyond the rapid. He had a vision of where they were going.

In the military, achieving objectives is just as important as it is in business. However, as someone once said, "No soldier ever charged a machine gun emplacement because it was part of his annual objectives." Machine gun emplacements are charged frequently, however, sometimes at great cost. Military leaders are forced to decide that they need charging, and they are able to motivate others to do it with them. They have a vision of what needs to happen and the ability to lead the charge.

They acquire that ability in part by the way the military trains leaders. In general, the military takes the youngest soldiers, trains them in tactics, makes them squad or platoon leaders, and then sends them out on a series of exercise missions during which they are increasingly responsible for achieving the objectives of the mission and safeguarding the lives of the people in their group. On these missions, they begin to test and develop leadership characteristics. If they are successful, they are given increased leadership responsibility as company and squadron commanders.

After they prove themselves as operational leaders, they are promoted to field grade, where as majors and lieutenant colonels they train and support the operational units. Senior management ranks—the colonels and generals—function at the highest levels, where they marshal resources and get them to the right spot at the right time so that the leaders at the unit level can do their jobs.

The military knows that leadership must occur at all levels of the organization, and people at all levels need the chance to develop leadership through experience and increasing responsibility.

> The late, great coach of the Green Bay Packers, Vince Lombardi, had a vision for his team: "The Green Bay Packers never lose—sometimes the clock runs out." The Packers played every game that way.
>
> In the early '60s, the Packers played for the championship against the Dallas Cowboys at Green Bay. It was below zero. The game went back and forth, and it was very low scoring. Finally, the Packers were down by more than a field goal but less than a touchdown. In the last minute of the game, they had driven to the Cowboys' goal line. The Packers needed the touchdown. Twice the Cowboys stopped them on the goal line. The third time, Jerry Kramer threw a block, and Bart Starr sneaked through for the touchdown. The Packers won the game and the championship.
>
> In the locker room afterward, Kramer—the offensive lineman who made the winning block—said, with tears streaming down his face: "We love each other; we love our coach. Being here is sort of like being in Camelot."[1]

How do great leaders such as Vince Lombardi make people feel like they could never lose? He communicated a vision to his team, and he motivated them to fulfill it because they shared the vision. The story in the sidebar shows what happened if a player did not share the vision. Sometimes when people do not share the vision, the leader needs to deal with the situation. That is one of the toughest jobs leaders have—to separate the organization from people who do not share the vision. Lombardi knew it.

One year, during contract negotiations, a player came into Lombardi's office. Remember, this was the late '50s and early '60s, before all the lawyers and agents. Players negotiated their own contracts. This player said, "Vince, I'd like you to talk to my lawyer." Lombardi said, "What?" The player said, "I really don't feel I can do this alone. I need you to talk to my lawyer. He's my agent. You need to negotiate with him." Lombardi said, "Excuse me just a minute," and he left the room. He came back 5 minutes later and said, "Tell your lawyer to negotiate with the Pittsburgh Steelers; you've just been traded."[1]

A few years ago, Japan Airlines had what was the worst airline disaster in its history. Many people were killed. The next day, the CEO said: "It's my job to run an airline where that kind of thing cannot possibly happen. I failed. I resign." And he did so. In Japanese society a few years earlier, he might have taken a different, more drastic course of action. Either way, the sense of personal responsibility is clear. How many leaders would take things that seriously?

Moses is another example of a good leader, especially because he is proof that leaders do not have to do everything themselves; they just need to cause change to happen. Moses was educated. He was rash—he made a couple of moves early in his life that could have gone against him. He was scared. He ran away at one point in his life for 40 years, lived in the desert, got married, and raised a family. Finally, the vision came, and he knew he had to return and communicate the vision. Moses could not do it himself, however, because he stuttered. His brother, Aaron, did the talking for him.

That should be comforting. A lot of people who talk about what they need to do to become leaders often say, "I'm not charismatic; I'm not good at being a cheerleader." That's all right, because what they really need to do is lead; not all leaders are charismatic. Moses was not. He was able to communicate his vision through his brother. He motivated people by getting them to see his vision and by instilling some discipline in them.

However, Moses was a very poor manager. At one point, when the Jews were wandering through the desert during their 40-year journey,

Moses's father-in-law, Jethro, realized that Moses was spending all his time dealing with day-to-day issues between people. The vision and the mission were getting lost in petty squabbles.

Jethro told Moses, "You need to adopt some administrative procedures so the organization can function smoothly on routine matters. Then you need to appoint yourself some administrators—managers, directors, vice presidents—to oversee the day-to-day operations."

Of course, Jethro did not exactly say that. He said, "Thou shalt teach them ordinances and laws and shalt show them the way wherein they must walk and the work that they must do. Thou shalt provide able men to be rulers of thousands, rulers of hundreds, rulers of fifties, and rulers of tens. Every great matter they shall bring to thee, but every small matter they shall judge."

Jethro knew that there needed to be a system to keep the organization going, even in times of tremendous change. Jethro was the first FQM consultant.

What about our leaders today? Three excellent examples are Don Peterson of Ford Motor Company, David Kearns of Xerox, and Roger Milliken of Milliken & Company.

At Ford 12 years ago, Peterson stood up in a room of his senior executives and said, "I am the problem. I just realized it. I intend to change....I intend to change the way I act. Any of you want to come along, come along. It's going to be fun. Any of you don't, tell me right now." Over the last 12 years, we have seen what has happened to Ford. By all measures, the company is an astounding success.

> "I am the problem. I just realized it. I intend to change....I intend to change the way I act. Any of you want to come along, come along....Any of you don't, tell me right now."
>
> Don Peterson
> Ford Motor Company

In the last 10 to 12 years, Kearns has led Xerox to a tremendous turnaround in its business fortune. Kearns said, "Who says we can't have zero defects? Who says we can't have 100% outgoing quality. Not me, not you. Let's get it done."

At Milliken & Company about 8 or 10 years ago, one of the managers challenged Milliken during a meeting. The manager said, "You know, of all the people in this room, there are only three willing to listen." Milliken jumped up on one of the little banquet chairs, raised his right hand, and said, "All of you repeat after me: I will listen. I will not shoot the messenger. I realize that management is the problem." Then they started to change the way they ran the company.

Well-led enterprises share the following characteristics:

- An organizational vision that is clearly defined at all levels
- The ability to communicate that vision
- Members motivated to fulfill the vision
- Shared cultural values or norms that exemplify the vision

Sometimes, even in the best-led organizations, instead of a single leader, there is a group of leaders. Sometimes, leadership just happens among people who have been chosen or among whom it has fallen to lead the organization.

No one individual led the American Revolution. Still, a single vision guided it: "We hold these truths to be self-evident: that all men are created equal." The multiple leaders of the Revolution were able to communicate that vision.

Eisenhower said, "Leadership is the art of getting people to do what you want done, when you want it done, in the way you want it done, because they want to do it!" That is not an easy task. A leader is the single person who has his or her back turned to everyone else. That is a dangerous position; a leader must know where he or she is going and that people will follow.

Dr. Galen Barbour, Associate Chief Medical Director for Quality Management for the Department of Veterans Affairs, tells managers that they must remember: "It isn't people that cause problems, it is the process, so don't waste time pointing fingers....If you are going to fix the process, the people in the process must do it....Change must be data driven; measure and remeasure to see if you are achieving your goals."

Dr. Barbour sums it up in one powerful statement: "You have to lead *and* get out of the way!"

A successful quality effort must have both leadership and vision. Leadership must step up to the challenge. When that happens, the quality process will greatly enhance the ability of an organization to meet its objectives.

INDIVIDUAL RESISTANCE AND SKILLS DEVELOPMENT

While the ability to motivate others to believe in the vision is a key to successful change, employees who do not have the skills necessary to deal with the change will react with great anxiety. Resistance to change in those circumstances is all too common. Yet when people are properly prepared for change and believe they have the skills to make it happen, they will participate with motivation and enthusiasm.

To unleash these forces of executive leadership, shared vision, and the power of employee participation, managers need to create an environment in which people are motivated and equipped to translate the vision and values into their daily behavior. In such an environment, employees are actively involved in making decisions and implementing changes that help the organization achieve its goals. Those changes improve customer focus and satisfaction, reduce cycle times, increase productivity, lower costs, increase return on investment, and ensure high-quality products or services.

> When visiting a major hotel chain's property, the enthusiasm of the staff was obvious. A front desk clerk wore a large button that said "I'm empowered." When asked what that meant, she said, "I can do anything to keep a customer happy so long as I don't give away the property." She was asked how and replied, "If something goes wrong, I can give you one of these green tickets for a free drink or meal and if needed even a room." However, when asked whether she ever looked at the processes to avoid having to compensate for mistakes, she answered with a worried look, "What's a process?"

Organizations do not achieve this kind of employee involvement without a focused and highly defined strategy for developing employee skills. To achieve these results, management must provide the necessary skills development to realize the vision, mission, and business objectives of the organization.

In order for FQM to succeed, people and organizations need both clearly defined reasons for the needed change and the ability to make it happen. Management must provide that focus. Therefore, as the discussion on *planning* points out, management must charter, align, and *train* cross-functional teams to accomplish the things required to satisfy the critical success factors and achieve the strategic objectives of the organization.

Employee teams often react with confusion, anxiety, and frustration when given process improvement projects that they could not

Process improvements are being made all over the globe in a wide variety of industries. For example, Western Geophysical, a leading seismic exploration company, achieved great success drilling seismic test holes in the mountains of Colombia. A joint team of company, customer, and supplier personnel chartered by Gary Jones, Vice President for Latin America, increased daily output from two holes a day to five holes a day while raising the success rate from 75% to more than 95%. This improved capability won them new work because quality was most important to the customer. These results show that people listened when Orval Brannan, President of Western Geophysical said, "Quality must become an integral part of every plan and every project, an integral part of the thinking of each employee."

accomplish. For example, some time ago, an aerospace firm asked for outside advice on finding the reasons for its failure to realize results from its TQM efforts. The firm's management felt that if its teams could just hear a pep talk on the values of continuous improvement, they would begin to be effective.

The firm had already spent $500,000 on TQM training for all employees. Then, 12 months later, management assigned employees to 66 teams and instructed them to create projects for organizational improvement. By the time the firm asked for advice, 64 of the teams had disbanded without achieving any sustained results. The two remaining teams were still trying to define their charters. It was too late for help. TQM had failed.

How could the quality effort fail after the company spent so much money and time on training? The message was clear: employees will not develop skills and confidence from training unless they can apply it to projects that are clearly focused on achieving organizational goals.

Employees need management leadership and support, quality tools and process analysis, and skills training to focus their process improvement efforts. Only then are the chances for measurable success high. Employees need to experience success—and management needs to recognize that success—with process improvements that they believe make a difference. To do that, they need training. Without training, they will fail; with it, the teams will succeed. Their success solidifies the skills and confidence necessary to make process improvement in the way of life in the organization.

INCENTIVES

Before the organization can change, people must change their behavior. To effect this, incentives must be either created or modified. For example, people who will share the benefits of an improvement (gain sharing) are much more motivated to change than those who are not able to share. Gain sharing is one of the incentives rated highest; in addition, it is free! In effect, managers agree to share something that does not yet exist as a means to achieve a simple and smart decision.

Ranked just behind gain sharing as incentives for change are two more free items: saying "thank you" and a visit from management. "Thank you" says exactly what it seems to: it tells the person who made the change happen that it is appreciated. The power of thanking someone for his or her support, in writing or orally, is impressive.

The high value placed on a visit from management may seem a bit surprising to some. A line worker explains, "When I see management in my work area it means what I am doing is important to them." He then added, "That means I am important. I like that feeling!" That attitude is true of other workers as well. Even if management visits to criticize a less than desirable outcome, the message is still clear—what workers are doing is important.

> **O**ne Air Force General pointed out the value of a personal visit by saying "When I go to a unit, I always stop by the battery shop. It is always fun for two reasons. First, I get to watch the colonel try to find it. Second, the look on the young airman's face is always a delight. You see, that young airman has a dirty, dangerous job and my visit lets him know that I care. That's important to him and to me."

An evaluation of existing incentives should reveal whether or not they reward the cross-functional team-based approach. In most cases the answer will be no. Most reward systems focus on individuals, rather than the team. That has to be changed, and the best way to do so is to put an employee team to work on improving it. Remember, reward and recognition is a key enabler for FQM.

RESOURCES

An organization must commit adequate resources—both people and money—to achieve success with process improvement. Otherwise, instead of leading the charge, the organization will fall prey to anxiety and failure.

Six activities, in particular, require resources. Those in the *prepare* phase are

1. Management training in concepts
2. Assessment
3. Management guidance

One in the *plan* phase is

4. Organizational awareness training

Those in the *deploy* phase are

5. Team training in concepts and tools of quality
6. Team participation

To alleviate some of the immediate drain on resources, many organizations use consultants. Of course, in the long run, the organization needs to own the change process and be capable of sustaining the effort without a great deal of outside assistance.

The resources required will vary according to the size of the organization and the scope of the change process. The best advice to organizations that ask what resources to commit is to think in Juran's terms.

According to Juran, an organization basically has two management tasks. One is to manage the day-to-day operations of the organization. That task controls processes, and change is not a desirable element here. The other task is to improve or reengineer processes so that performance increases. Most organizations spend little time on the latter task. An organization should spend at least **10%** of its time on improving process performance.

It is true that at the outset, management in particular experiences an increased work load. However, after the processes are improved, and a shift in culture takes effect, managers spend less time fighting fires and have more time to carry out true management functions.

In the beginning, most organizations will experience strain in terms of both time and money in starting the change process. Training takes time and money. Within a year, however, most organizations will benefit enough from process improvements that they will be effectively able to run the day-to-day operations of the business with 90% or less of their resources. This frees up the time, talent, and money needed to improve. Process improvement will then have come a long way toward being an institutionalized part of the management process.

Capability gap analysis is one analytical approach many businesses have used very effectively to find resources. An organization should analyze its processes in terms of how important they are to its critical success factors versus how well the organization performs them. Next, the organization should identify the processes that it is doing well but which are not important to its strategic objectives. Then, the organization should determine how to transfer resources used in doing the relatively unimportant things so that they can be used to improve the important processes that are not being done well.

Most of the time, though, leadership resolve is the key ingredient in making committed resources available. John Micham, CEO of AT&T Paradyne, is a good case in point:

> Micham brought his 40 key managers together to train in quality awareness and process requirement concepts. He dubbed them the "40 fanatics," and they committed with him to change the company. He then selected 6 key executives from different functions of the business and dedicated their efforts, as a team, to leading the change effort.

One year later, after a good start and early successes, the executives resumed their key positions in the business. AT&T Paradyne survived its problems, and today the company continues to implement process improvement as an institutionalized part of its management process. There was never a doubt about management's willingness to commit the resources or about the need for change and increased performance.

Remember, committing resources to support the change process is a business decision. The costs of process improvement are an investment in the business. Therefore, a measurable return on investment can be seen as a result of those costs. Organizations can achieve significant measurable results within six months. Over time, overall gains can range from two to ten times the dollars invested.

ACTION PLANNING

After lack of leadership commitment, the most next frequently encountered barrier to successful implementation of quality FQM and process improvement initiatives is lack of a structured and well-managed action plan. Without that, teams will suffer false starts, which undermines credibility and confidence by playing into the hands of skeptics, who in most cases are already saying: "We tried that before and it didn't work."

Many organizations take a serendipitous approach to quality improvement. They trust to luck, saying, "Let's get everybody trained and stay out of their way; good things will happen." Nothing could be further from the truth. Without action planning, nothing good happens. Time and money are squandered, and morale and confidence evaporate. This approach to quality is ineffective.

Three factors are required for successfully planning and implementing FQM activities:

- ▪ Management direction
- ▪ Structured implementation
- ▪ Monitoring process

Management Direction

In this chapter, the difference between leadership and management has been stressed. Although leadership was said to be a key factor, management skills are required for successful action planning and implementation of FQM.

It is management's job to provide focus and set the direction for change. Management must understand the organizational vision and determine the critical success factors that will make that vision real. Management must also decide on the process strategies to achieve the critical success factors. Management must define the cross-functional FQM projects that will make the process improvement strategies effective. These project teams must have a clear charter from management so that they can understand the focus, scope, measurements, and goals of their activities. Only then do FQM teams have a real opportunity to produce lasting results for the organization.

People generally do not like to be involved in the change process without a clear understanding of what is expected of them and how they can meet those expectations. Ambiguous expectations produce anxiety, confusion, and poor results.

One example of effective management involvement in action planning is Louisiana Offshore Oil Platform, Inc. (LOOP), which is in the business of unloading crude oil from tankers and pumping it via an 18-mile underwater pipeline into underground reservoirs. The crude oil is stored in the reservoirs until needed by customer refineries. Then, the crude oil is pumped via a pipeline to the refinery.

Although the company's president, Bill Thompson, felt that the company was doing well overall, he decided that the company needed significant improvement to continue to meet its growth goals. He had experimented with Total Quality Management training and process teams without dramatic success. Then he heard about the approach described in this book, and he liked what he heard. He had a problem, though: how to convince his board to fund FQM efforts given the current lack of any substantial financial results.

Given this situation, he and his outside consultants decided to work on getting some relatively short-term financial results and then let his management team and people prove the value of the process. The consultants worked with the management team to understand the vision and growth objectives of the company. They identified critical success factors specifically targeted at those objectives.

One critical success factor identified was higher utilization of the pipeline. This factor was discussed at some length. It was decided that it was a marketing issue. The only way to produce higher utilization was to sell more contracts; therefore, it was not a likely project for a FQM team.

At this point, the management team was asked if there were not other possibilities for a FQM project focusing on the unloading, storage, and delivery processes. What about efficiency? "No problem," management said. "We only lose 1.25 barrels of crude oil per 1000 barrels in the unloading and storing process." "Doesn't that cost money," the consultants asked, "and isn't the customer dissatisfied that you lost the crude oil?" "Not really," they said. "You see, the customer allows us to lose (keep) one barrel per thousand in the process as part of their quality tolerances. Therefore, we're only stuck with the cost of 0.25 barrels per thousand."

The next question asked should have been obvious, but evidently it had not been focused on before: "What would it mean to you if you had the capability to lose zero crude oil in the unloading process?" Bill

Jennings, the director of operations, did some fast calculations. He looked up with excitement on his face and said, "We would save $4 million per year, direct to the bottom line."

More discussions ensued, and a FQM project team of the appropriate technical, operational, and administrative personnel was put together and chartered to pursue this goal. Three months later, the team recommended adding some highly accurate and tightly calibrated measuring technology and adjusting some other operational procedures. In two months, the crude oil loss had been cut in half—saving the company $2 million per year. That achievement will also protect the company if customers tighten their acceptance tolerances for lost crude oil.

Without this company's management focus, specific FQM team improvement efforts would probably not have lasted. Needless to say, the board has continued to fund the FQM efforts. The team members are excited and proud of their accomplishments and the recognition they have received from management.

An executive said, after learning about Focused Quality Management, "It seems like it's just common sense." He then smiled and said, "But common sense wouldn't be so valuable if it was common, would it?"

In summary, it is the job of management to focus and direct cross-functional FQM teams. To do this, management must translate the vision into customer-focused, value-added strategies and projects that lead to improved organizational performance. Being responsible, management must stimulate employee self-directed actions and manage the change process in a way that fosters employee ownership. Management, as the controller of most resources, is responsible for providing the systems, the skills development, the incentives, the resources, and the detailed plans to ensure that successful change takes place.

BIBLIOGRAPHY

1. Kramer, Jerry, *Instant Replay*, New American Library, 1968.

8

IS THE
ORGANIZATION READY?

No matter how eager you may be to start making Focused Quality Management (FQM) work, some questions need to be answered before beginning.

It is the organization's leader who must be able to decide if the organization is ready. If the answer is no, any initiatives that are undertaken are sure to fail.

If the leader of the organization can answer yes to the questions that follow, the organization is ready to proceed. If not, the pertinent steps in the *prepare–plan–deploy–transition* process should be reviewed and the appropriate actions taken.

DOES THE ORGANIZATION HAVE A VISION?

If the answer is yes, try to write down the vision from memory or at least describe what it means. If the leader of the organization cannot describe the vision, then it is necessary to return to Chapter 2 and begin again.

A good test of how well the vision has been internalized by those who will be trying to achieve it is to ask two or three employees what the vision is and what it means. If they cannot describe it, then Chapter 5 should be reviewed again for ways to improve the communication plan. It's not working!

If the vision, as well as communication and implementation of it, are effective, everyone in the organization clearly understands it. People must be able to translate it into day-to-day actions that affect what makes the vision a reality. The job of the leader is to clearly make the link between what people do, where the organization hopes to go, and what it wants to be.

For example, one mutual fund firm was able to make the link between a vision that depicted "a responsive customer-focused organization" and a mailroom employee. The employee revised his mail-sorting process and break schedule to ensure that customer inquiries reached a responsible party in hours rather than the days it had taken using the old process. The results were a motivated employee, a better process, and, more importantly, satisfied customers.

HAVE THE MISSION AND VALUES OF THE ORGANIZATION BEEN DEVELOPED AND CLEARLY COMMUNICATED?

Ask people what they think the purpose of the organization is. Determine if there is a corporate ethic that reflects shared values.

If employees can explain the company mission and values in their own words and know how they are supposed to act as they pursue the goals of the organization, then there is a solid foundation to build on toward achieving the strategic goals. If they cannot, then the company must start over and correct what went wrong in the first pass at implementing FQM.

HAS AN ORGANIZATIONAL ASSESSMENT BEEN DONE?

Part of building the base for successful FQM is an assessment. That assessment should target the key things the organization must do to satisfy its customers. Then, the processes that have been identified must be measured and analyzed to determine how well (or poorly) they are performing.

For example, a hotel may observe check-in to determine how long the process takes, how many times guests get to the right room without any problem, and the accuracy of the reservations system. A manufacturing firm may measure manufacturing cycle time, the percentage of defects, and the cost of scrap. A retailer may track stock-outs and returns. The important thing is to focus on what is important to the customer.

Managers should know how the processes are performing. If they do not, the company should go back to square one (Chapter 2). If the performance of the key processes has not been assessed, then managers are not prepared to effectively focus the quality management initiative. Time and money need to be invested to find out where the organization is now. The expense pays dividends later, when the company is better equipped to set a course that gets it where it wants to go.

During the assessment of key processes, the survey in Appendix A, *Quick Quality Survey,* can be distributed to a cross-section of the organization to identify the differing views among people at various levels and in various functions. This can even be done after an assessment has been completed. The varied responses and the strikingly different viewpoints that people have based on misunderstandings of top management's intentions, expectations, and actions may be enlightening.

A word of caution: *Don't overreact!* Even if the survey and assessment show that the company is not even close to where it wants to be, there is no need for alarm. Instead of a quick, easy answer (which is usually wrong), a better remedy is to follow the step-by-step planning process detailed in Chapter 3.

HAVE STRATEGIC BUSINESS GOALS BEEN IDENTIFIED?

In crafting the statements of vision, mission, and values, the strategic goals that need to be achieved in order for the vision to become a reality should have been identified. For example, a strategic goal for an organization that is to become a global competitor might be to have a worldwide presence in target markets. A strategic goal for an organization that is to be the technological leader in its industry might be to achieve a dominant position in creating and patenting leading-edge, innovative, breakthrough, technical solutions. In reality, strategic business goals are nothing more than tightly focused mission statements.

The most effective way to prepare a list of strategic business goals is for top management and the senior leadership team to each draw up separate lists of goals based on each group's understanding of the company vision statement. The two lists are then consolidated into one master list that includes the goals agreed upon by both groups.

Then, the strategic business goals must be further broken down into critical success factors. As explained in Chapters 2 and 3, critical success factors are the absolute essentials, the *musts*.

HAVE THE KEY BUSINESS PROCESSES BEEN IDENTIFIED?

Once the critical success factors have been clearly stated, it is time for top management and the senior leadership team to identify key business processes. *Key business processes* means the things that the organization does to survive. For example:

Type of organization	Key business process
An aerospace firm	Design rocket motors
An engineering firm	Create drawings
A toy company	Market toys
Every organization	Hire and train people
A mutual fund firm	Invest money
Every organization	Manage resources

The possibilities are endless; however, any company should be able to identify the relatively few (eight to ten) broadly stated processes that describe what it does. Once the key business processes have been listed, the managers are ready to move on and measure the impact of these processes on the critical success factors.

HOW DO KEY BUSINESS PROCESSES AFFECT CRITICAL SUCCESS FACTORS?

Impact analysis is easy enough to do. As discussed in Chapter 2, it consists of a simple question: What is the impact of this particular process on the critical success factors? Is it relatively low, medium, or high?

The process prioritization matrix in Chapter 2 (Figure 2.6) is a handy tool for summarizing the findings. (A duplicate of the matrix is provided in Appendix B.)

Note that some processes may affect certain critical success factors but not others. Other processes may have little or no impact on any critical success factors. In the latter case, there is waste in the organization. Whatever the impact analysis reveals, the important thing is to clearly establish which processes have the most influence on the factors critical to success.

HAVE THE CAPABILITY GAPS BEEN IDENTIFIED?

After considering impacts, the next step (as explained in Chapter 3) is to determine what gaps exist between the ideal level and the current level of performance for each key business process. The assessments done during the *prepare* phase should provide the information necessary to make those judgments.

Once the gaps have been measured, they can be combined with the scores from the impact analysis. This will indicate which of the key business processes are the strongest performers and which are the weakest. The weakest ones are the best targets for process improvement. Identifying the targets for improvement is a significant step toward focusing quality management. As discussed in Chapter 4, the gaps can be closed by tapping into the reservoir of knowledge, dedication, and enthusiasm offered by the people involved in the process.

DO EMPLOYEES WANT TO DO A GOOD JOB? DO THEY HAVE WHAT THEY NEED TO DO SO?

The answer to the first question is almost always yes. Too often, however, the answer to the second is no.

Why? Because in most organizations, the people closest to actually producing goods and services are regarded as doers, not thinkers. The doers are expected to simply do what they are told to do. As a result, they never get the information, tools, or authority they need to make real decisions. Lacking that support, they often do things that work against rather than toward the ultimate goal.

> In the 1980s, a military education facility proudly unveiled a new motto: "Think or die." One former commandant, seeing it for the first time, said, "I kind of hoped there were more choices!"

Sadly, when the final undesirable outcome results, the people in the process often say that they thought what they were doing made little sense, but they did not have the authority to deviate from the instructions they received.

In many ways, this situation is similar to the lines from *The Charge of the Light Brigade*, by Alfred Lord Tennyson:

> *Theirs not to reason why,*
> *Theirs but to do and die.*

Such sentiments sound heroic, but in most cases, they are just not smart. Recall Tennyson's next two lines:

> *Into the valley of Death*
> *Rode the six hundred.*

A much better choice is to share the vision and the things identified as essential to achieving it. At the same time, it should be made clear that it is everyone's job to make the vision a reality by *thinking* about and *doing* those things that make it happen.

To help the *thinking*, employees must be trained in the tools of process analysis. Some managers take the stance that workers have trouble using process analysis tools or resist getting involved in planning and managing the process. Those managers are wrong on both counts.

People have little difficulty describing, flowcharting, measuring, and analyzing work processes. Why should they? They are the ones who do the job every day—often for years and many for decades. Rather than resist involvement, they welcome it and the opportunity to shape the future.

The late Coach Ray Elliot of the University of Illinois often used a story to illustrate the consequences of over-controlling people's options. He told of a head coach who, with little time left in a tied football game, told his quarterback: "Listen carefully and do what I say. Go in there, run the ball off tackle for three plays, and then punt."

The young man did exactly as he was told. The first play gained 25 yards, from the 20-yard line to the 45. The second went from the 45 to the opponent's 30-yard line. The third took the ball to the 12. On the next play, the quarterback punted the ball.

As the quarterback came back to the sidelines, the irate coach said, "What in the world were you thinking out there?" The young man replied, "I was thinking, boy, do we have a dumb coach."

Given a clear vision of the strategic goals of the organization, most workers demonstrate a surprising (to many managers) ability to

The general manager of a resort hotel in Maui reported that he had yet to find anything his staff could not do once he freed them to get involved and improve the way they served the customer. He also reported that other general managers who had not yet empowered their staffs were amazed at the apparent success of his approach.

make tough decisions about staffing, elimination of processes, and all the determinants of organizational success or failure. They show that they quickly understand the tools of process analysis, but can also effectively apply them to improve their own work processes. In a word, when it comes to their own jobs, the workers are the *experts*.

IS QUALITY REWARDED?

A study of manufacturing workers in one plant showed that while gain sharing ranked first, a personal thanks was the next most valued reward.

Managers too often underestimate the effect of their personal thanks. An aerospace CEO reported that he was stunned when he attended the wedding reception of a worker's daughter and saw framed on the wall of his home a handwritten note of thanks that he had sent years earlier. Realizing the impact of his simple gesture, he resolved to say thank you more often.

Once the process improvement teams have completed their initial activities, it is time for celebration. This shows that the organization welcomes initiative, innovation, and involvement.

Whether or not it is fashionable to admit, people enjoy feedback for their efforts. Feedback from a leader carries a great deal of weight. People usually strive to please the boss.

When management appreciates something, it should show. The people involved should be thanked on their own turf, preferably in the presence of their co-workers. The word will spread quickly that the boss rewards quality.

IS THE ORGANIZATION FOCUSED?

This final question is the key to success. As pointed out in Chapter 1, the primary reason why quality initiatives fail is due to lack of focus on the critical success factors that are essential to achieving business objectives. After the vision has been created, the factors critical to successfully achieving the business objectives should have been clearly identified during the *prepare* and *plan* phases. These critical success factors are impacted by how well the key processes that most seriously affect each of the factors are performed. The assessment revealed how well or poorly those key processes are performed. During the *plan* phase, the key processes that were most important but which lacked adequate levels of performance were identified. The improvement initiative must be focused on those specific processes or it will surely fail.

It should be possible to trace each project initiated back to the vision, using the following guide:

Project Focus Guide

The goal of the _____ process improve-
(team name)

ment team is to improve the _____
(process or subprocess)

process. _____ are to be _____ by
(Desired outcomes)   (increased/reduced)

_____ in _____ . These results will enhance
(percent or amount)   (time frame)

achievement of _____ , which
(specific critical success factor)

directly impacts the _____ .
(vision/segment)

If the team goals can be traced back to specific critical success factors and objectives that have an impact on the vision, it is reasonably certain that the organization is focused for success.

IS THE ORGANIZATION READY?

The answers to the questions in this chapter should determine if an organization is ready to start and follow through on a process improvement initiative. Planning, commitment, and follow-up should guarantee success, which will be worth the effort.

9

WHEN THINGS GO WRONG

If you have started your process improvement initiative, but nothing seems to be happening, read on. Possible solutions are provided in this chapter.

THE WHOLE IS GREATER THAN THE SUM OF THE PARTS

The key to success in process improvement is how effectively change is managed. The components of managing change are depicted in Figure 9.1

The five parts of the equation—vision, skills, incentives, resources, and action plan—add up to change. If even one part of the equation is missing, all efforts are doomed to failure.

The rest of this chapter provides a series of questions that can be used as a checklist to determine whether process improvement efforts are on track or are veering off course.

HAS THE VISION BEEN COMMUNICATED?

Vision is essential because it lets people know the direction intended for the company. Vision sets the goal so that, even without being

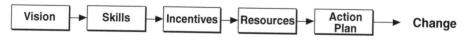

FIGURE 9.1

FIGURE 9.2

monitored, people know what to do. Additionally, middle-level managers need to buy into the vision so that they, too, can effectively adapt to the changes that will be required of them.

Without vision, as depicted in Figure 9.2, people with the required skills, who are motivated to make improvements and have adequate resources and a plan, have no idea why they are being asked to change.

The employees of one firm expressed their confusion as follows: "I think we could do anything the boss wants, if we only knew what he wanted." The boss listened and crafted a clear vision that led to the first profits in a number of years. That is the power of vision.

If people and the process seem confused, a review of Chapter 2 should help in getting the vision communicated and understood. When the vision is clear to everyone in the organization, tangible results will be the outcome.

DO EMPLOYEES HAVE *ALL* THE SKILLS THEY NEED?

Skills give people the ability to realize the vision. The seven quality problem-solving tools help people to improve their process analysis skills:

- Flowchart
- Cause-and-effect diagram
- Pareto chart
- Histogram
- Scatter diagram
- Run charts
- Control chart

All are essential. For example, those charged with improving a key business process must be able to construct a process flowchart. They also must understand cause-and-effect diagrams. In order to accurately measure effects and isolate root causes, they must be able to collect and analyze data.

Interpersonal skills as well as implementation skills are also very important. Finally, at the worker level, core competency job-related skills are essential for improved job performance.

In addition, at the supervisory and middle management levels, the following skills are key to successful implementation:

- Team building
- Team leadership
- Coaching
- Facilitation

At the highest levels of the organization, skills in the following areas are required to effectively focus management activities:

- Strategic planning
- The seven management tools
- Process tracking through effective measurement

Leave out skills, as in Figure 9.3, and people who know what is required of them, are motivated to make it happen, and have adequate resources and a plan do not know how to make it happen. As a result, anxiety sets in.

In a large service organization, people were put on process improvement teams without training in the quality tools. They were so distraught at their inability to make the required improvements that they simply refused to participate. Senior management then assigned a consultant to help the teams, train the people needing tools training, and get the process moving again. It worked, but only after some needlessly anxious moments for the team members and management.

If employees are negative and worried, perhaps their training in understanding and using quality tools was not thorough or broad enough. Perhaps not every team member was trained, or those who

FIGURE 9.3

were trained have forgotten what they learned. Any of these situations can cause unnecessary and unproductive anxiety. Whatever the reason, it must be addressed, because anxiety is contagious.

HAVE PEOPLE BEEN GIVEN INCENTIVES TO SUCCEED?

Incentives must be aimed at encouraging people to shed functional perspectives, work as teams, and strive for continuous process improvement. Team incentives based on cross-functional, customer-centered, strategically focused results are the hallmarks of most successful improvement initiatives. They are absolute necessities for motivating people to change.

If little has been seen in the way of results, the reward and recognition system has probably not been changed (Figure 9.4). The existing system was designed to reward the old way of doing things, and it is probably doing just that. As a result, however, the change currently envisioned will not come about completely or quickly.

For example, Focused Quality Management calls for cross-functional teams working to meet customer needs. However, most existing systems reward individual and functional performance. The organization wants one thing, but rewards the opposite. The answer is team recognition, group rewards, cross-functional incentives, and management support of the behaviors it wants to encourage.

In the 1970s, the U.S. Air Force was trying to mold an effective team, but at the same time it instituted a competitive officer-rating system. The system allocated top ratings (1) and certain promotions to 22% of those rated. Unfortunately, the system also mandated low ratings (3) and almost certain nonpromotion to 50% of those rated.

In a system where nonpromotion was grounds for terminating a career, the results were predictable. Teamwork disappeared. Officers were pitted against their peers, and the organization suffered.

A favorite Pentagon story circulated about the young captain who

FIGURE 9.4

asked a fellow captain to work late on a hot project. The reply said it all: "I'm sorry, you must have me confused with a 1. I'm a 3, and 3's don't work late or on weekends." Fortunately, the Air Force soon replaced the system.

If improvement is moving too slowly or is motivating people to do something but not necessarily the right thing, comparing the current reward and recognition system with the one proposed in Chapter 5 may be beneficial.

DO PEOPLE HAVE ADEQUATE RESOURCES?

Resources are just as critical. People need time to focus on improvement. Regularly scheduled and held team meetings must be part of the process. Funds (usually not significant) for testing proposed improvements and securing needed assets should be available to the teams. Moreover, the team charters should clearly delineate the authority to call on others for support when needed.

People who are making jokes about the process improvement initiative instead of making haste to accomplish it may be expressing frustration. Frustration is the natural outcome when people are ready and willing to create change but lack the resources to bring it about (Figure 9.5).

If adequate resources are not provided, people may know what is expected of them, have the skills to make it happen, and be motivated to do it, but they cannot hope to succeed. It takes only a short time for reality to smother the flames of enthusiasm so carefully fanned.

People react with anger to canceled team meetings, delays, and back-pedaling. The amount of anger and frustration is directly proportional to how well the vision, training, and motivation have succeeded. Therefore, the process must be monitored closely to ensure that meetings are not overcome by competing priorities and that penny-wise and pound-foolish decisions do not stifle the initiatives.

FIGURE 9.5

Process improvement must focus on results; however, as the improvement initiative begins, it is often necessary to track activities to make sure that the necessary things are really being done and that adequate resources are being devoted to the effort.

An effective technique is to track the number of canceled meetings. Another method is to track attendance to find out if any part of the organization is not freeing up people to participate in the process. The best way is for managers to visit team meetings and see for themselves if the team has adequate resources. This sends a signal throughout the organization: this initiative is important. That is usually all that it takes.

HAS THE ORGANIZATION PLANNED FOR SUCCESS?

Last, but clearly not least, a detailed, specific action plan must be in place. At the very least, for each part of the improvement process it should include:

■ Schedules of activities

■ Review milestones

■ Completion dates

Action without a plan is much worse than a plan without action. A plan without action is a waste, but fortunately few resources are involved (generally just the planners), and the waste is not widely visible. Action without a plan is usually much more costly in terms of squandered resources, time, and, most important, credibility.

Without a plan, the step-by-step process so necessary to creating significant and lasting change cannot and will not happen. Instead of change, chaos is the likely result; in place of success, failure; in place of commitment, contempt. All this results from false starts (Figure 9.6).

False starts are double-edged swords. First, they waste resources. Second, they destroy credibility by confirming the opinions of the pessimists and cynics within the organization. It is like taking two steps

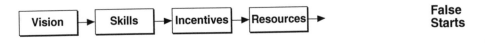

FIGURE 9.6

forward and three steps back. Valuable enthusiasm is squandered as people grumble, "I told you it wouldn't work here!"

At that point, the action plan should be reviewed to find out if people are using it or determine if it is still the right approach. Chapter 3 may provide some assistance in this case.

ARE YOU FOCUSED TO SUCCEED?

Answer this question by using the Project Focus Guide in Chapter 8 to verify that your organization is targeting its improvement efforts on the "vital few." If the important processes are not being targeted, the organization is wasting resources and enthusiasm.

It would be ideal if everything worked well from the start, but sometimes even the best laid plans go astray. You can be sure of that. The evidence, however, is clear: Focused Quality Management works. It works in hospitals, manufacturing plants, hotels, banks, flying squadrons, coal mines, government agencies, and even the IRS. It works because management leads a motivated, trained work force to make improvement a way of life. It works because it is focused, it is quality, and it is management. That is its competitive edge.

Appendix A

QUICK QUALITY SURVEY

Name: _____   Company Name: _____

Position: _____   Company Business (Main): _____

Perceived Quality
Awareness/Knowledge: (*) _____   Size Of Company: _____

Please answer the following questions about your company. The questionnaire is designed to be completed within 20 minutes; it is, therefore, very subjective and is seeking your immediate perceptions. Rate each question on a scale of 1 - 10, 1 being low and 10 being high.

Scale (*):

1	2	3	4	5	6	7	8	9	10
Low				Medium					High

	Assessment 1	Assessment 2
1. How involved is the leadership in promoting quality?	☐	☐
2. How effectively is the quality vision and values communicated and re-enforced to all employees?	☐	☐
3. How appropriately are responsibilities for quality improvement assigned at top management level?	☐	☐
4. How appropriate is the scope and depth of information for supporting continuous quality improvement?	☐	☐
5. How effectively is the available information used for quality improvement?	☐	☐
6. To what extent is your company aware of the quality levels of its competitors or other organizations?	☐	☐
7. To what extent is planning for quality improvement performed at top management level?	☐	☐
8. To what extent is planning for quality improvement performed at middle management level?	☐	☐
9. How effectively are quality plans resourced, reviewed and steps taken to ensure accomplishment?	☐	☐

Quick Quality Survey (continued)

Scale (*):

1	2	3	4	5	6	7	8	9	10
Low				Medium					High

	Assessment 1	Assessment 2
10. To what extent is the Human Resource plan derived from and integrated with the Quality Plan?	☐	☐
11. To what extent do you think employees of all categories are: a. involved in quality planning	☐	☐
b. involved in working on improvement teams	☐	☐
c. encouraged to improve/innovate	☐	☐
d. trained in quality improvement skills	☐	☐
e. recognized for quality improvement	☐	☐
12. To what extent does the company assess employee attitudes/motivation, etc.	☐	☐
13. Overall, how well do you think that the company manages Human Resources, across all its dimensions (training, rewards & recognition, Health & Safety, etc.)?	☐	☐
14. To what extent are customer requirements the prime focus in planning the processes for producing products and services?	☐	☐
15. How well is quality controlled in the processes which directly produce goods and services?	☐	☐
16. To what extent are variations in process performance analyzed with a view to improvement?	☐	☐
17. To what extent are suppliers included in quality improvement?	☐	☐
18. To what extent are the supporting services (e.g., marketing, accounting, administrative services, etc.) involved in improving their processes in order to meet or exceed their "customers'" requirements?	☐	☐

Quick Quality Survey (continued)

<u>Scale (*):</u>

1	2	3	4	5	6	7	8	9	10
Low				Medium					High

	Assessment 1	Assessment 2
19. To what extent does the company track quality improvement/trends in: a. product and services processes	☐	☐
b. supporting services	☐	☐
c. suppliers	☐	☐
20. To what extent do these trends indicate continuous improvement: a. product and services processes	☐	☐
b. supporting services	☐	☐
c. suppliers	☐	☐
21. How effectively is the current and potential customer base analyzed to determine the differing requirements/expectations of each segment?	☐	☐
22. How effectively identified are the product and service features for each requirement/expectation?	☐	☐
23. How effectively is a "customer focus" attitude (external and internal customers) promoted throughout the Company?	☐	☐
24. How effectively are customer relationships managed (e.g., access, follow-up, training of customer contact personnel, complaint analysis, etc.)?	☐	☐
25. How comprehensive and well defined are standards for customer contact (e.g., responsiveness, accuracy of information, complaint resolution)?	☐	☐
26. How effective are the methods used for determining customer satisfaction?	☐	☐

Quick Quality Survey (continued)

<u>Scale (*):</u>

1	2	3	4	5	6	7	8	9	10
Low				Medium					High

	Assessment 1	Assessment 2
27. To what extent are customer satisfaction indicators trended?	▢	▢
28. To what extent are favorable trends being achieved in levels of customer satisfaction?	▢	▢
29. To what extent are comparisons made with competitors providing similar products/services?	▢	▢
30. To what extent is total customer satisfaction regarded as the principal strategic objective?	▢	▢

"Quick" Quality Assessment Summary

Category	Question #'s Relating to Category	A Total # of Questions	B Total Points: Assessment 1	C Total Points: Assessment 2	D Average per Category Assessment 1 (B ÷ A)	E Average per Category Assessment 2 (C ÷ A)	F "Baldrige" Conversion Factor	G "Baldrige" Score Assessment 1 (D × F)	H "Baldrige" Score Assessment 2 (E × F)	Differences + or – (G – H)
Leadership	1–3	3					10			
Information and Analysis	4–5	2					7			
Strategic Planning	6–9	4					6			
Human Resource Utilization	10–13	*8					14			
Quality Assurance	14–18	5					18			
Results	19–20	*6					15			
Customer Satisfaction	21–30	10					30			
							TOTALS:			

* Includes Sub-questions

Appendix B

CSF/KEY PROCESS MATRIX

Process Prioritization Matrix

	Critical Success Factors						Total Impact	Process Performance	Process Performance Gap	Weighted Gap	Priority
Key Processes											
1											
2											
3											
4											
5											
6											
7											
8											
9											
10											

Rating Key:

Process Impact on CSF:
1 = Low
2 = Medium
3 = High

Process Performance:
1 = Inadequate
5 = OK
9 = Very Well

175

Appendix **C**

EVOLUTION OF QUALITY MANAGEMENT

While the 20th century has been the century of productivity,
the 21st century will be the century of quality.

J.M. Juran

To understand why so many quality efforts fail and how Focused Quality Management (FQM) can help get them back on track, it is necessary to understand how the quality movement has evolved in this country. Many organizations have tried to force-fit quality improvement efforts into a variety of management styles that originated as far back as the Industrial Revolution. In doing so, they have ignored fundamental steps—such as **involving upper management and improving key business processes to support strategic goals**—even though they have been proven time and again to work. Organizations that want to improve their business processes but attempt to do so without changing their functional structures, despite the proven success of cross-functional teams, guarantee their own lack of success.

During the last two centuries, management has concentrated on productivity—how best to use machines and how best to motivate employees. Typically, organizations have decided that there is one best way to pursue productivity. For these organizations, management techniques are an all-or-nothing approach; there is only one right way to manage, whether the organization has adopted a mechanistic or a humanistic mindset.

Organizations with mechanistic mindsets have tended to center on efficiency alone. Functional departments produce their products or services in an isolated, get-it-done-fast manner. No one really worries about quality until the end-of-the-line inspection. Individual departments might ensure quality in their own areas, but **no one in the organization institutes quality across the board**.

> "**W**e're the inspection department, and our job is to look at these things after they're made and find the bad ones. Making it right in the first place is the job of the production department."
>
> J.M. Juran[1]
> "What Japan Taught Us about Quality"
> *The Washington Post*

Organizations with humanistic mindsets pursue quality through a host of approaches—from participatory management and quality circles to self-directed work teams and quality-of-work-life initiatives. **For all the flurry of activity, however, there has been precious little in the way of tangible results.** Here again, the problem seems to stem from the isolated, sporadic nature of such activities. Few organizations are looking at the big picture. Few organizations are institutionalizing quality as a pervasive way of life throughout the organization. In effect, the approaches have created islands of excellence in an ocean of opportunity.

Even so, the quality consciousness of the last few decades has set the stage for Juran's "century of quality." To show how, the effects of a century of productivity since the Industrial Revolution are traced in the following sections.

NINETEENTH-CENTURY INDUSTRIALISM

By the late 1800s, the United States had become the largest and most competitive industrial nation in the world. To deal with their new machines, factory owners began to make major changes in the design and control of work. Division of labor, as put forth by Adam Smith in his 1776 work *The Wealth of Nations,* became the standard operating procedure. Manufacturers sought to increase productivity by subordinating workers to machines and supervisors.

Not surprisingly, machine-oriented management approaches evolved to guide managers as they ran the organization. The most common were

- ▪ Bureaucratic
- ▪ Classical
- ▪ Scientific

Bureaucratic management embraced the ideas of Max Weber in an attempt to manage through administration of the organization. Weber espoused precision, speed, clarity, regularity, reliability, and efficiency. The way to attain those ideals was through strict division of labor, hierarchical supervision, and detailed rules and regulations. Every gear had its wheels; every wheel had its cogs. When meshed properly, the result would be a smoothly running machine.

Classical management built on the same precepts, but took them one step further to encompass the design of the total organization. This philosophy maintained that management's responsibilities included planning, organizing, directing, coordinating, and controlling the organization's work through its functional departments.

Finally, scientific management, mostly guided by Frederick Taylor, centered on designing and managing individual jobs. Taylor held that human performance could be defined and controlled through work standards and rules. He advocated the use of time-and-motion studies to break down jobs into simple, separate steps to be performed over and over again without deviation. His focus was on finding the "one best way" to accomplish any task.

> "**W**e will win and you will lose. You cannot do anything about it because your failure is an internal disease. Your companies are based on Taylor's principles. Worse, your heads are Taylorized, too. You firmly believe that sound management means executives on one side and workers on the other. On the one side, men who think and on the other men who can only work. For you, management is the art of smoothly transferring the executive's ideas to the worker's hands."[2]
>
> Konosuke Matsushita
> Founder and Executive Director
> Matsushita Electronics Industry
> Osaka, Japan

Thus, scientific management, or Taylorism, adopted the attitudes of its time; that is, treat the organization like a machine. Workers should fit the requirements of the machine-defined job. Managers should plan, organize, direct, coordinate, and control. All thinking should be done by the managers and all producing by the workers. The thrust was strict subordination, rules, and regulations. Organizations were supposed to be rational and operate in the most efficient manner possible.

From the late 1800s to the end of World War II, the mechanistic philosophy dominated organizational life. Although lacking in humanism, it nonetheless reflected the realities of the times. The United States was being flooded with immigrants seeking a new life and willing to work hard. Mechanism offered an efficient way of employing this largely uneducated, unskilled work force.

Mechanism also suited the conditions of the Great Depression, when people desperate for employment cared little about working conditions. The combination of available machine technology and an abundant work force produced management systems that would capitalize on both, making the United States the preeminent economic force in the world.

> "**U**ntouched by war, the industrial heartland churned out cars, washing machines, vacuum cleaners, mixers, lawn mowers, refrigerators, stoves, furniture, carpets, all the appurtenances for the mushrooming postwar suburbs...Unfortunately, quality in the post World War II years took a back seat to production—getting the numbers out. Quality control came to mean end-of-the-line inspection. If there were defects and rework, there would be profits enough to cover them."
>
> Mary Walton[3]
> *The Deming Management Philosophy*

U.S. market dominance also got an assist from the outcome of World War II. As Europe and Japan concentrated on rebuilding their war-torn economies, the United States returned to the peacetime production of consumer goods. Embracing mechanistic principles and techniques, manufacturers rallied to the call of the richest generation in American history: Give us goods, goods, and more goods. Quantity, not quality, was the goal, and American business met it.

Mechanism made the fortune of the United States as a nation. However, it also produced the vertical, functionally focused organizational structure that has proven to be a major deterrent to improving cross-functional business processes.

TWENTIETH-CENTURY HUMANISM

As U.S. productivity was hitting its stride, concern about negative effects on workers began to gain momentum. With the rise of the human relations movement, the pendulum started to swing away from caring about the omnipotent machine and toward caring about people.

The human relations movement focused on the relationship between productivity and satisfied workers in cohesive work groups. The theory advocated organizing workers into self-regulating groups. Managers were told to foster positive work relationships and treat workers well. The result, according to this theory, would be that group members would support high production goals, solve problems, and help one another as needed. Simply put, if small groups of workers were given maximum freedom to control their work, the business would prosper.

By the 1950s, however, it was clear that the simple solutions proffered by the human relations movement could not guarantee productivity. In fact, research showed that unhappy workers in isolation could be productive and that satisfied workers in cohesive groups could be unproductive.[4]

In the 1950s and 1960s, behavioral scientists began to play a major role in business schools, basing their teachings on theories of human motivation. Techniques like Theory X vs. Theory Y and participatory management resulted. The behavioral scientists pointed to worker motivation as the key to productivity.

Simultaneously, socio-technical systems emerged. Such systems aimed to bring the pendulum back to center, balancing the social needs of the organization with its technical needs. Socio-tech's developer, Eric Trist, described it as being brilliantly simple: For an organization to operate effectively, its technological system (machines) must mesh with its social system (people).[4]

More recent approaches have included quality-oriented, people-focused initiatives, such as self-directed work teams, quality of work life, and job enrichment, to name a few. Such approaches have not performed as well as hoped. For one thing, questions have arisen about how to balance control and decision-making between managers and workers. For another, these approaches have been criticized as too often emphasizing the worker's needs and too seldom producing positive bottom-line

> "The performance improvement efforts of many companies have as much impact on operational and financial results as a ceremonial rain dance has on the weather...This rain dance is the ardent pursuit of activities that sound good, look good, and allow managers to feel good—but in fact contribute little or nothing to bottom-line results."
>
> Robert Shaffer and Harvey Thomson[5]
> "Successful Change Programs Begin with Results"
> *Harvard Business Review*

results. Because they take so many forms, these activities have been called "flavor of the month" programs.

These approaches and techniques are not inherently bad. They can be valuable tools. However, to be most effective, they must be integrated with and support a quality improvement process that is focused on process improvements to meet the strategic goals of the organization.

THE EMERGENCE OF TQM

Although many people believe that TQM got its start in Japan, history credits its inception to work done in the early 1900s to improve agricultural efforts in Britain and railroad management in the United States. TQM really started to take form, however, when Walter Shewhart pioneered what came to be known as statistical quality control (SQC), at AT&T Bell Laboratories. As originally conceived, SQC was intended to apply statistical techniques to examine variations in manufacturing processes. Shewhart's first memo on the subject was dated May 16, 1924.

The two founding fathers of TQM, W. Edwards Deming and J.M. Juran, worked for Bell Labs during the 1920s and were intimately involved in developing and testing Shewhart's statistical concepts. During World War II, Deming taught SQC to U.S. companies engaged in wartime production. It was the means by which American industry produced the wide range of interchangeable parts needed to stock the Allied arsenals.

"Japan paid dearly for its participation in World War II. Of its major cities, only Kyoto had escaped wide-scale damage from aerial bombings, and 668,000 civilians had died. The nation's industrial base was in ruins; agricultural production was off by a third...the once prosperous populace had gone first without consumer goods, then without food for the wartime effort. By 1947, there was little of either."

Mary Walton[3]
The Deming Management Method

After World War II, however, U.S. industries lost interest, regarding SQC as time-consuming and unnecessary. Working frantically to put out an unprecedented volume of goods, workers felt that SQC hardly mattered. Any defects could easily be made up in quantity. The results proved them right: large quantities of defective products were produced, but the market still demanded more.

Japan, however, was not faring as well. All too aware that Japan was facing a life-or-death economic crisis, the Union of Japanese Scientists and Engineers asked Deming to help its members increase the country's productivity.

In a series of eight day-long lectures, Deming introduced SQC to Japan. Eventually, Deming broadened his view from the purely statistical pursuit of quality to a list of 14 points, or quality objectives, that constitute the philosophical underpinnings of TQM as it is known today. In his view, managers were the key to quality.

According to Karou Ishikawa, a leader in the Japanese quality effort, all was not smooth sailing at first. In *What Is Total Quality Control?*, Ishikawa says that the Japanese quality movement made limited progress in the years immediately following Deming's first visit. Ishikawa credits Juran with helping the Japanese understand that quality is everyone's concern, from managers to clerks.[6]

In the 1940s, Juran pointed out that the technical aspects of quality control had been well covered, but companies did not know how to manage for quality. He identified some of the problems as organization, communication, and coordination of functions—in other words, the human element. According to Juran:

> An understanding of the human situations associated with the job will go far to solve the technical ones; in fact, such understanding may be a prerequisite of a solution.[7]

The results of the Japanese TQM efforts of yesterday are obvious in today's world markets. Given the Japanese success, all nations recognize that their businesses must follow suit if they are to survive in a global economy.

This widespread recognition is a direct result of the efforts of Deming and Juran. Until his death at age 93 in December 1993, Deming continued to lecture and consult worldwide. At 87, Juran is hard at work on a history of quality management. In a recent interview, he was asked about the future of TQM in this country and abroad:

> Right now, three major forces are putting pressure on TQM and will most influence its future direction.
>
> One is multi-national, global competition, which is already evident and requires no explanation.
>
> The second is technology. I sometimes describe it as life behind the quality dikes. Just as the Dutch pushed the sea back

from the shore to develop property, they've had to pay an enormous price by building and maintaining the dikes. And if something goes wrong, the consequences are also enormous. So it is with our dependence on technology. We can't afford poor quality. Look what happened at Chernobyl.

Finally, the ecology is poised to become the dominant influence in the future.[8]

Juran's last point, about the ecology, was addressed by *Business Week* in a special edition entitled "The Quality Imperative":

Traditionally, companies have reacted to environmental rules by installing technology to control pollutants. But as the legal liabilities escalate and environmental laws get tougher, this firefighting approach looks antiquated and expensive. Quality management, which preaches prevention and continuous improvement, gets you out of the reactive mode.[9]

It is clear that the United States is emerging from a century of productivity, as suggested by Juran. Undeniably, the country has benefitted economically from that focus. Nevertheless, like many good things, the old ways of managing appear to have come to the end of their useful lives. As evidenced in the United States, the emphasis on productivity without quality has led to many organizational failures.

In a recent *Washington Post* article, Juran described how one of the most financially powerful companies in the United States was taken completely by surprise and nearly paid the ultimate price:

In the 1950s and 1960s, Xerox had a lock on a key industrial process—copying documents. Everybody wanted Xerox copies, and nobody could get them except by leasing Xerox machines. The company was growing, and Xerox executives could look at their instruments and see sales, costs, and profits at a glance. But they had no meter showing customer satisfaction.

The Xerox machines malfunctioned or broke down regularly, and Xerox executives knew it. They could have sent their designers back to the drawing board to redesign the machines so they wouldn't fail. Instead, they created a service force they could dispatch to fix the machines. As far as Xerox executives were concerned, that solved the problem.

Xerox's customers didn't agree. They didn't want repairs, they wanted machines that didn't break down in the first place....

The senior management lacked summarized information on field failures and their effect on customer relations, the performance of competing machines and the extent of customer defections.

The Xerox scenario has become a familiar one in the United States. In the 1950s, there were about 30 U.S.-owned companies making color televisions in the United States. By the 1980s, we were down to one.[1]

As business and industry reaped the rewards of a century of productivity, it also appears that they sowed the seeds of quality and competitive failure. What can be done to regain the competitive edge? The answer is Focused Quality Management.

BIBLIOGRAPHY

1. Juran, J.M., "What Japan Taught Us about Quality," *The Washington Post*, pp. H1 and H6, Aug. 15, 1993.
2. Townsend, Patrick L. and Joan E. Gebhardt, "A Quality Problem," *Quality*, pp. 101–102, March 1989.
3. Walton, Mary, *The Deming Management Method*, New York: Perigee Books, 1986.
4. Sashkin, Marshall, "An Overview of Ten Management and Organizational Theorists," *University Associates*, pp. 206–221, 1981.
5. Schaffer, Robert H. and Harvey A. Thomson, "Successful Change Programs Begin with Results," *Harvard Business Review*, pp. 80–89, Jan.–Feb. 1991.
6. Kaoru, Ishikawa (translated by David J. Lu), *What Is Total Quality Control?—The Japanese Way*, Englewood Cliffs, N.J.: Prentice-Hall, 1985.
7. Gagne, James, "Quality Performance Means More at Dow," *Dow Chemical Company, CPI Purchasing*, 11 pp., March 1986.
8. Juran, Joseph J., personal communications, May 17, 1993.
9. Smith, Emily T., "Doing It for Mother Earth," *Business Week*, pp. 44–49, Fall 1991 (bonus issue).

INDEX